A theory of the super soldier

MANCHESTER
1824

Manchester University Press

A theory of the super soldier

The morality of capacity-increasing technologies
in the military

Jean-François Caron

Manchester University Press

Published by Manchester University Press
Altrincham Street, Manchester M1 7JA
www.manchesteruniversitypress.co.uk

British Library Cataloguing-in-Publication Data
A catalogue record for this book is available from the British Library

ISBN   978 1 5261 1777 9   hardback
ISBN   978 1 5261 4364 8   paperback

First published by Manchester University Press in hardback 2018
This edition published 2019

The publisher has no responsibility for the persistence or accuracy of URLs for any external or third-party internet websites referred to in this book, and does not guarantee that any content on such websites is, or will remain, accurate or appropriate.

Typeset
by Toppan Best-set Premedia Limited

Для Венеры и Мари-Габриэллы
Венера мен Мари-Габриэллаға
For Venera and Marie-Gabrielle
Pour Venera et Marie-Gabrielle

Contents

Acknowledgements

I would first like to thank Nazarbayev University for providing me with the opportunity to present the main findings of this book to various conferences throughout the world. Arguments developed in this book also benefited from discussions with my colleagues Philip Enns and Spencer Willardson, as well as the comments from the anonymous reviewers, whose suggestions have all been highly appreciated. A word of gratitude should also be extended to Brandon Brock, who generously agreed to review some parts of the manuscript. I also wish to thank all the team at Manchester University Press for their high level of professionalism, which helped to make the whole process smooth and efficient.

Above all, I wish to thank my wife and my mother-in-law for supporting me in the last couple of months and for allowing me the necessary time to write this book. Because of that, they have spent more than their fair share of time with our newborn daughter. I promise I'll make it up to you. Paxmet!

Introduction

Since the beginning of history, we also have wanted to become more than human, to become Homo superior. From the godlike command of Gilgamesh, to the lofty ambitions of Icarus, to the preternatural strength of Beowulf, to the mythical skills of Shaolin monks, and to various shamans and shapeshifters throughout the world's cultural history, we have dreamt – and still dream – of transforming ourselves to overcome our all-too-human limitations.

Fritz Allhoff et al., 2010

When most people think of super soldiers, they usually refer to comic book characters, such as Captain America and Iron Man, or to fictional book and movie heroes like Jason Bourne. These fictional soldiers are the result of the military's desire to create the perfect combatant with superhuman physical features (increased endurance, focus, and pain threshold) or state of mind (elimination of fear, stress, or fatigue). While these examples relate to contemporary fictional characters, we must acknowledge that the desire to transform soldiers into super combatants is not a mere fantasy. On the contrary, it has always been an explicit objective of the military institution. History provides us with numerous examples of the various attempts made by different countries in this regard. Of course, some examples are more grotesque than others, such as the use of paranormal techniques to practise mind reading, remote viewing, and other tactics that were promoted by Lieutenant Colonel John Channon, who – as reported in Jon Ronson's book *The Men Who Stare at Goats* (2004) – wanted to create the 'First Earth Battalion' as a way to develop non-destructive methods of conflict resolution. However, many other historical instances were more down to earth than this far-fetched example, such as the spear and shield of the Greek hoplite or the armour worn by medieval knights. Alongside these mechanical technologies, armies have also resorted to various forms of chemical intervention, such as improved techniques of fighting that are not restricted to better weapons or equipment, but include drugs or medicines that affect the cognitive capacities of combatants. For example, Homer's *Odyssey* refers to the use of opium by the Greeks as a way to reduce the grief and sorrow associated with the loss of comrades as well as to combat stress and other forms of war trauma (Kamienski, 2016, p. 32). Inca warriors were known for using coca leaves to enable them to fight fatigue and to increase their resistance to pain (Kamienski,

2016, p. 46), while the Zulu warriors used traditional plants with psychotropic effects when they fought against the British in the nineteenth century. As described by Lukasz Kamienski, the impact of such chemical enhancements was enormous:

> The Zulus fought with fanaticism, dedication, and fury. Armed and fortified by their shaman doctors with potent intoxicants, they went into battle utterly without fear. … Even when injured they did not stop fighting because their bodies were rendered insensitive to pain through the use of powerful anesthetizing plant remedies. The Zulu warriors seemed immune to the enemy rifle fire, so they readily launched almost suicidal massed charges and incredibly easily retained their combat effectiveness. (Kamienski, 2016, p. 86)

It is also well known that many soldiers who fought in the Second World War had very easy access to hard drugs. Such was the case with the Finns who between 1939 and 1940 used massive quantities of heroin, morphine, and opium during their fight against the Soviet invaders (Kamienski, 2016, pp. 132–140), who, for their part, resorted to 'trench cocktail', a mixture of vodka and cocaine. However, the most famous example undoubtedly remains that of the Nazis, who patented Pervitin (the ancestor of crystal meth) and distributed it in industrial quantities to members of the *Wehrmacht* after observing its propensity to induce increased vigilance, resistance to fatigue, and a sense of invincibility. For the Nazis, this drug was seen as a necessary tool in their quest to enlarge their vital space. Its main supporter, Professor Otto Ranke, who chaired the Berlin Institute of Physiology, believed that Germany had one enemy that was more powerful than the Russians, the French, and the British combined: its soldiers' tiredness. However, this strange and elusive enemy was considered a contingency that could be overcome through the help of this little pill, which contributed to the initial successes of the German army. Indeed, the 35 million tablets ordered by the *Wehrmacht* before the campaign in France are now considered to have been one of the factors that contributed to the rapid victory of Germany over the Anglo-French troops in the spring of 1940 (Ohler, 2016). In fact, as Peter Steinkamp, a historian of medicine, notes, 'the *Blitzkrieg* was only possible because of methamphetamine. In fact, it was founded on the use of this drug.'[1] The effects of Pervitin also caught the German High Command off guard, leading to the fabrication of the myth of Erwin Rommel as one of the fastest tank commanders of the war. Fuelled by Pervitin, the future legendary 'Desert Fox' knew no limits. Indeed, even his corps commander, General Hermann Hoth, was unable to reach him as his orders always arrived when Rommel was already leading his men further into the enemy's territory (Ohler, 2016, p. 88): a situation that led Rommel's 7th *Panzerdivision* to be renamed the *Gespensterdivision* (Ghost Division).

In 1944, the Nazis developed a more powerful drug called D-IX, a composite of oxycodone, cocaine, and methamphetamine that was intended for special submarine commandos who had to stay awake during an entire mission of up to four days (Paterson, 2006). The Allies responded with their own field tests on the effects of amphetamines on soldiers' performance and allowed their soldiers to use Benzedrine, a mixture of amphetamines developed in the 1930s that increased

.soldiers' confidence and alertness, though to a lesser degree than Pervitin. During the course of the war, 140 million tablets were supplied to Great Britain and to the US armed forces. The use of amphetamines did not stop with the end of the Second World War. These drugs were also widely used by both sides during the Korean War and in other subsequent South East Asian conflicts; as reported by a member of the US Air Force, these pills were available 'like candy' during the Vietnam War (Cornum, Caldwell and Cornum, 1997). Nowadays, the use of dextroamphetamine (known as a 'go-pill') remains standard during fatigue-inducing mission profiles, such as night-time bombing missions.

To a large extent, while some of these drugs are still used today, they nonetheless belong to a bygone era. Indeed, the current research devoted to increasing the physical capabilities of soldiers sounds much more like attempts to transform members of the military into comic book superheroes. The development of an exoskeleton by the Defense Advanced Research Projects Agency (DARPA), which has been responsible for the technical innovations of the US Army, is a clear example of this type of innovation. According to DARPA, this exoskeleton will allow a soldier to carry 45 kg of equipment while reducing his metabolic consumption by 25 per cent, and it should be operational in 2018.[2] The French army is making similar strides: the Directorate General of Armaments (DGA) introduced the Hercule exoskeleton in 2011 to enable soldiers to carry loads of 100 kg with little or no effort.[3] DARPA has also admitted that it is trying to develop a technology called Z-Man that will enable soldiers to climb walls in a manner similar to a gecko.[4] During a demonstration in 2014, an individual weighing 100 kg was able to climb a glass wall eight metres high while effortlessly carrying a load of 25 kg thanks to a simple pair of gloves.[5]

However, military research is not solely limited to increasing the physical capabilities of soldiers through devices that are extrinsic to the human body. It is also actively involved in developing technologies and drugs with the goal of altering – sometimes permanently – the internal physical faculties of individuals as well as their cognitive abilities. Alongside ancient forms of neuropharmacology, new ways – which are sometimes as far-fetched as the one of John Channon – are being studied in order to reach this goal. Numerous armies are now aiming to increase combatants' cognition and their capacities to learn and train as well as developing human–machine interfaces to ameliorate their psychological and physical weaknesses. In many respects, these science-fiction-esque developments raised the prospect of a 'Human Enhancement Revolution' (Savulescu and Bostrom, 2009; Allhoff et al., 2010). In fact, several reports mention research that is consistent with this approach, which has been described by the bioethicist Jonathan Moreno as 'the fastest-growing area of science' (Hanlon, 2011). We need only to think of research designed to erase certain events from the memory of soldiers to prevent post-traumatic stress disorder (Lehrer, 2012), or whose objective is to change the cellular and genetic structure to enable them to run longer distances, to survive longer without food, or to be able to consume foods that are not normally digestible (such as grass) (Shachtman, 2007), to erase pain with the help of a vaccine, or to develop ways to stop bleeding with the wave of a wand (Knefel, 2016). Although this quest might seem unrealistic, it is nonetheless part of DARPA's mandate, which is, according to one of its high-ranking officials, 'about trying to do those things, which are thought to be impossible, and finding ways to make them happen' (Moreno, 2012, p. 26). For its part, the British Ministry of Defence has also launched a series of similar initiatives. For instance, a consultation paper published in 2010 by the Secretary of State for Defence states:

> Knowledge about the human brain is rapidly increasing including: understanding pharmacological effects to enhance performance and using brain activity to control systems. As such, it offers significant opportunities for defence and security in understanding adversaries' behaviours, training and improving human performance on the battlefield or in human-based security situations such as guarding or search. (Secretary of State for Defence, 2010, p. 24)

Consequently, the British government has recently started to provide funding in various areas, such as robust and fieldable techniques for neurological imaging, bio-electronics integration, and programmes aiming to exploit the subconscious (Royal Society, 2012, p. 7).

Thus we are probably about to enter a new paradigm as the wars of tomorrow run the risk of being carried out by 'super soldiers' with physical and cognitive capabilities that currently belong to the world of science fiction and comic books. This possibility, which is becoming increasingly real and inevitable but surprisingly remains neglected by ethicists, opens the door to a series of fundamental questions: are the motives behind the development of these technologies by the military-industrial complex noble or are they simply a way to sacrifice soldiers' health for the sake of military efficiency? Are all these enhancement technologies morally problematic? If not, by what criteria might we be able to sort the ones that are acceptable from the ones that are not? Will these innovations breach the moral principles of 'Just War'? What are the possible legal implications of the use of these technologies and this medical research? What should the ethical parameters of military research be? These are the burning issues into which this book will delve, because technological discoveries that do not take into account their consequences on humankind ruin the human soul (Rabelais, 2006). In order to avoid the potential consequences of these types of capacity-increasing technology, there is a need to understand them strictly through the lens of ethics. After all, if ethicists remain so very silent about this hubristic technological trend, human societies are at risk of downfall if they do not pay attention to the dangers associated with a willingness to go beyond their natural limitations.

However, one mistake would be to fall into the trap of arguing strictly in favour or against these new developments in warfare, as if it were self-evident that this question could be solved in binary terms. When analysed from an ethical viewpoint, the development and use of capacity-increasing technologies in the military is far more complex than it first appears because it presents us with a significant moral dilemma. On the one hand, enhancing soldiers' capacities can be interpreted as a moral obligation on the part of the military; on the other hand, such technologies may also end up contravening fundamental moral principles of warfare. Therefore, any analysis on this question needs to be nuanced. While it is necessary to be critical of such military enhancement, it would be a mistake to condemn these studies as being entirely evil and inadmissible.

This nuanced evaluation is largely dependent on an interaction between four moral premises. On the one hand, like all other professional organizations, the military (1) has a moral duty to protect its members' lives and health as much as possible by allowing them to fulfil their duties with minimal risks. In return (2), these technologies should be morally acceptable only insofar as they respect a Kantian deontological ethics, which implies that the military has a universal non-contingent moral imperative to never perform an act that would result in using its members as pawns or tools; in addition, (3) they must not run counter to the principles of Just War Theory or (4) alter the modern principle of equality between human beings. The reader should be aware that this ethical stance will be central throughout this book.

This study will be divided into five chapters. First and foremost, the investigation of the ethical character of military research cannot be performed without a conceptual clarification of the various studies conducted by the military-industrial complex. More specifically, it is important to distinguish between methods aiming to restore the physical capabilities of soldiers from those looking to increase them. Although capacity-restoring technologies are not exempt from certain questions, those raised by capacity-increasing technologies are far more problematic from an ethical perspective, precisely because the latter tend to provide unnatural advantages to their beneficiaries: a situation that contradicts a moral pillar of our modern societies. There therefore is a need to understand the moral implications of this inequality in the military world and to determine whether the inequality inherent to the use of capacity-increasing technologies is immoral. As it will be argued, despite the fact that soldiers benefiting from capacity-increasing technologies will undoubtedly have an important edge over their enemies who cannot enjoy such methods, this advantage does not challenge in any way the nature of the ethical principles surrounding the morality of warfare, which do not see military asymmetry as a morally perverse situation to be avoided and condemned.

The second chapter will highlight the inherent moral necessity of these technologies as a corollary of the military's obligation to protect its members and as a potential way to increase the morality of warfare. In fact, their development is intimately linked with this desire on the part of the military institution to respect its duty of care towards its members. Secondly, it will also argue that capacity-increasing technologies can play a significant role in the enhancement of morality in warfare. Indeed, despite the best intentions and means deployed by the military institution, soldiers very often remain the weakest link when it comes to respect for the laws of war. Recent history has shown that warfare causes psychological trauma that can contribute to transforming the best-trained combatants into murderous agents unable to distinguish between legitimate and illegitimate targets. In this sense, we cannot ignore the possibility that certain capacity-increasing technologies might play a vital role in respect for the moral rules of war by thwarting this killing instinct.

However, despite the moral obligation to use capacity-increasing technologies, their development and use should also be evaluated under the light of the Kantian deontological notion that presupposes that such technologies should never treat soldiers as instruments and should not have a negative impact on the moral rules of warfare. For this reason the third chapter will oppose the intrinsic moral value of capacity-increasing techniques with these moral imperatives. Indeed, these research efforts pave the way for a crucial threat that cannot be ignored, namely, to harm soldiers' health, to deprive them of their obligation to disobey illegal and immoral orders, and to negatively affect the moral principles that address the fair termination phase of a war. As it will be argued, the use of technologies and medicines by members of the military is not only potentially dangerous to their physical integrity, but can also harm their moral agency. This latter possibility could create a situation in which certain soldiers would avoid the legal consequences of potential crimes they might commit. Through the use of very contemporary examples, this implication of capacity-increasing technologies will show that this possibility is not merely a hypothetical fantasy but a troubling reality that must be fully considered.

To prevent such a threat from being realized, the aforementioned risk will be followed by a discussion of the ways to chart the world of military research and its inherent problems. At first glance, it seems that the criteria that are used for civilian research should simply be applied to the military sphere. However, such a possibility turns out to be more problematic than it first appears, especially with regard to the notion of consent, which cannot be applied in the same way within the military as it would be in civilian research. If we seriously take into account the particularities of the military, it is important to reflect on the standards that should determine the ethical acceptability of capacity-increasing techniques and the boundaries that states should not cross in their quest to transform fiction into reality.

After discussing the current types of capacity-increasing technologies (both mechanical and chemical interventions), this book will conclude with a discussion of what is obviously a forthcoming way of enhancing warfighters' capacities: genetic manipulation. Specifically, some military studies are suggesting the prospect of transhumanism as a tangible reality.[6] This fear is animated by the fact that enhancement techniques might eventually increase soldiers' physical or cognitive faculties permanently through gene modification. Despite the fact that there are moral grounds justifying transhumanism, the use of gene therapy in the military faces significant ethical challenges that must be addressed, namely, the fourth moral premise presented earlier. Indeed, permanent capacity-increasing technologies that soldiers might end up acquiring during their employment in the armed forces may significantly alter the foundations of equality between human beings – a risk that cannot be ignored.

While it is true that many of us are afraid of technological progress, the fact remains that, just as when we face any other challenging question, unnuanced judgement based on fear should never prevail. Some social questions, such as this one about super soldiers, must not be discussed from a Manichaean perspective.

The best course of action is to rely on an impartial analysis organized around moral principles in order to sort the wheat from the chaff. It is hoped that this discussion will allow the reader to have a better and a more balanced understanding of the moral reasons that can justify the development and use of enhancement techniques within the military as well as the limits that such techniques should not exceed.

Notes

1 'Pervitin, la pilule de Goering', Arte Documentary, 2015, www.youtube.com/watch?v=1BHxWrZYlSI (last accessed 27 September 2017).
2 www.darpa.mil/Our_Work/BTO/Programs/Warrior_Web.aspx (last accessed 27 September 2017).
3 www.defense.gouv.fr/dga/mediatheque/videos/l-exosquelette-hercule (last accessed 27 September 2017).
4 www.darpa.mil/program/z-man (last accessed 27 September 2017).
5 /www.darpa.mil/news-events/2014–06–05 (last accessed 27 September 2017).
6 Many reports have suggested that some high-level athletes may already be using gene-doping, even though there have yet to be any confirmed cases of athletes genetically enhancing their bodies (Nilner, 2016; Friedman, 2014). This possibility looks credible at first sight, as leading scientists studying gene therapy claim to have been contacted by numerous athletes and coaches (Franks, 2014).

References

Allhoff, Fritz et al. 2010. 'Ethics on Human Enhancement: 25 Questions and Answers', *Studies in Ethics, Law and Technology*, Vol. 4, No. 1, pp. 1–39.

Cornum, Rhonda, John Caldwell, and Kory Cornum. 1997. 'Stimulant Use in Extended Flight Operations', *Airpower Journal*, Spring, pp. 53–58.

Franks, Tim. 2014. 'Gene Doping: Sport's Biggest Battle?', BBC, 12 January. www.bbc.com/news/magazine-25687002 (last accessed 27 September 2017).

Friedman, Lauren F. 2014. 'A Powerful and Undetectable New Kind of Doping may be Happening at the Olympics', *Business Insider*, 19 February.

Hanlon, Michael. 2011. '"Super Soldiers": The Quest for the Ultimate Human Killing Machine', *The Independent*, 17 November.

Kamienski, Lukasz. 2016. *Shooting Up: A Short History of Drugs and War*. Oxford: Oxford University Press.

Knefel, John. 2016. 'How the Pentagon is Building the Enhanced Super Soldiers of Tomorrow'. www.inverse.com/article/9988–how-the-pentagon-is-building-the-enhanced-super-soldiers-of-tomorrow (last accessed 27 September 2017).

Lehrer, Jonah. 2012. 'The Forgetting Pill Erases Painful Memories Forever', *Wired*, 17 February.

Moreno, Jonathan. 2012. *Mind Wars: Brain Science and the Military in the 21st Century*. New York: Bellevue Literary Press.

Nilner, Eric. 2016. 'Olympic Drug Cops will Scan for Genetically Modified Athletes', *Wired*, 28 July.

Ohler, Norman. 2016. *L'extase totale: le IIIᵉ Reich, les Allemands et la drogue*. Paris: La Découverte.

Paterson, Lawrence. 2006. *Weapons of Desperation: German Frogmen and Midget Submarines of World War II*. London: Chatham Publishing.

Rabelais, François. 2006. *Pantagruel*. Paris: Hachette.

Royal Society. 2012. *Brain Waves Module 3: Neuroscience, Conflict and Security*, February. London: The Royal Society Science Policy Centre.

Savulescu, Julian and Nick Bostrom. 2009. *Human Enhancement*. Oxford: Oxford University Press.

Secretary of State for Defence. 2010. *Equipment, Support, and Technology for UK Defence and Security: A Consultation Paper*, Cm 7989.

Shachtman, Noah. 2007. 'Kill Proof, Animal-esque Soldiers: DARPA Goal', *Wired*, 7 August.

1

Conceptual clarifications: distinguishing capacity-restoring and capacity-increasing technologies

It would be a truism to say that we currently live in a world where optimization is a dominant norm. We are indeed exposed on a daily basis to advertisements and undesired emails that are proposing to increase our muscular mass or our performance in many domains. The military world is no exception and is probably the sphere that, as the reader will soon realize, has taken this idea to its extremes, sometimes with terrible and harmful consequences for the soldiers. Nowadays, this trend has reached a science-fiction-like nature through the development and use of body armours and medicines that will soon allow service personnel to perform their duty in similar fashion to Iron Man, Batman, or Captain America.

However, before discussing the problems associated with the question of super soldiers, it is first necessary to establish an important conceptual clarification, so that the reader can understand precisely what this book is about and the moral issues it raises. We indeed need to understand what differentiates technologies aiming to restore individuals' physical or psychological faculties from those looking to increase them. While the former type of technology finds its moral justification in sound and convincing arguments, the latter are seen by certain authors as being more questionable. The main reason is undoubtedly the fact that they tend to provide their beneficiaries with advantages over others. For understandable reasons, our modern world, which is established on the notion that everyone is the bearer of unalienable and equal natural rights, has problems in coping with the prospect of capacity-increasing technologies. One consequence of this imbalance finds an echo in the military world, as the development and use of capacity-increasing technologies will, of course, end up creating two classes of combatants: those who can benefit from them and will consequently be less vulnerable to being wounded or killed, and those who do not have access to them and will consequently be more vulnerable to the deadly consequences of warfare. This imbalance between enemy combatants is seen by some authors as ethically problematic and as accordingly necessitating a negative stance regarding these technologies. As it will be argued in this chapter, this view is inaccurate. Of course, the military imbalance that results from the development of sophisticated technologies should not be ignored for certain weapons (especially drones), but capacity-increasing technologies are not ethically flawed, despite allowing some combatants greater chances than others of evading wounds or death on the battlefield. The primary reason is that they do not contribute to making their beneficiaries totally immune against lethal risks and, consequently, are not in contradiction to the way legitimate killing during warfare is morally established.

Capacity-restoring and capacity-increasing technologies in the military

It is necessary first to distinguish between two types of human optimization. As other authors who have investigated this question have noted (Lin, 2010; *Beyond Therapy*, 2003), we should not confuse enhancement with therapy.[1] The latter refers to treating someone who is suffering from a malfunction or a pathology that reduces her or his capacity to act like a normal human being, by giving her or him medicines or by using technology. As an example of this, we might mention of the use of Ritalin to treat attention deficit hyperactivity disorder (ADHD), which, according to the American Psychiatric Association, affects nearly 5 per cent of all children aged between four and seventeen. This disorder restricts the ability of those affected to become organized or to focus on tasks, and increases their tendency to make careless mistakes, not to follow social rules, to forget daily activities, and to be easily distracted by trivial noises or events, which, of course, seriously damage their chances to pursue their education at the same speed as their peers who are not affected by the disorder. This is why drugs, such as Ritalin, and special educational resources are seen as means to help them fight the associated problems and to not to be left behind other. In the same way, spectacles, laser eye surgery, and hearing implants to correct optical or auditory deficiencies, as well as replacement limbs and the use of medication to correct erectile dysfunction or to treat dwarfism with human growth hormones, can be seen as medical ways to restore someone's capacities to what is considered normal. Even sports organizations allow some of their athletes to use drugs in order to correcting medical problems that prevent them from performing at their full capacities. For instance, the International Cycling Union sometimes uses therapeutic use exemptions to grant athletes who are suffering from asthma or allergies the right to use substances that would normally be regarded as performance-enhancing agents. The famous Tour de France winners Christopher Froome and Bradley Wiggins have both been granted such exemptions, for use of the corticosteroid prednisolone to treat asthma in Froome's case, and of salbutamol, formoterol, budesonide, and triamcinolone acetonide to treat Wiggins's allergy to pollen.

On the other hand, enhancement aims to allow individuals to enjoy physical, psychological, and intellectual features or capacities that are beyond normal or beyond what nature gave them. The motivations that push individuals to use these methods are, of course, as extensive as the variety of human desires, but fact that we live in a society where performance and beauty are now dominating values seems to explain the willingness of many to increase their physical and psychological features (Ehrenberg, 1994).[2] The best-known examples of this are cosmetic surgery and doping in sports in order to improve an athlete's endurance and strength, as well as genetic manipulation to increase human lifespans or give new sensory-motor abilities.

Generally, technologies that have been primarily developed in order to restore people's capacities are diverted by individuals who wish to increase their performance beyond the normal. This is especially the case with students who are not suffering from ADHD but who nonetheless use Ritalin and related drugs before taking their Scholastic Aptitude Test (SAT) and other college admission examinations, as well as the use of Viagra or Cialis by men who are not suffering from impotence but who wish to enhance their sexual performance. Sports are probably the primary example of this misuse of capacity-restoring drugs. Nowadays, an almost endless list of athletes are caught for using human growth hormones or erythropoietin (EPO), a drug that was originally designed to treat people suffering from anaemia or inflammatory bowel disease and to increase the production of red blood cells in cancer patients.

In the world of the military, capacity-restoring technologies generally refer to devices for treating soldiers who have been wounded on the battlefield, by providing them with artificial limbs. For their part, capacity-increasing technologies used in the military refer to methods that enable soldiers to develop supernatural features. As mentioned at the beginning of the book, exoskeletons are a good example of these, as they allow the soldiers to carry enormous weights with very little effort. The same logic applies to the jetpack developed by DARPA (Higgins, 2014), which would make it possible for every soldier to accomplish with great ease what was considered, until it was achieved in 1954 by Roger Bannister, to be impossible for the human body: to run a mile in less than four minutes. Some myths indicate that people have been trying for over a thousand years to break this barrier, sometimes by using a form of 'encouragement' such as a bull behind them as an incentive to run faster. With the help of an exosuit-like device, the jetpack has shown that that this once unrealizable dream is now within the reach of anyone who is minimally physically fit. The development of the Z-Man technology follows the same path by allowing soldiers to climb walls like lizards while carrying a full combat load.

However, the military is also trying to increase the physical abilities of its service personnel through means that are intrinsic to the body. As mentioned previously, this quest has been pursued historically through the use of various drugs, such as cocaine and amphetamines. However the current trend in the military seems to have reached a point where these standard drugs, which are, of course, very dangerous, may become a thing of the past as new technologies and medicines become more sophisticated. For example, the pharmaceutical industry has recently developed an anxiolytic drug known as XBD173, which is deemed capable of eliminating fear immediately without causing addiction (Rupprecht et al., 2009). It has also been reported that DARPA is trying to develop technological devices that would allow the manipulation of soldiers' mental alertness and relieve stress. The organization has indeed funded a research programme at the School of Life Sciences at Arizona University to develop – in the words of the responsible researcher – ways of using transcranial pulse ultrasound to stimulate brain circuits (Tyler, 2010). The army could package this technology in soldiers' helmets, thereby allowing them to flip a switch that would enable them to fight stress and anxiety or boost alertness during long periods without sleep. In the light of the current development of pharmaceutical products, it seems that the military is also developing medicines that would be much more advanced and science-fiction-esque than traditional go-pills and other stimulants. Science has indeed developed a new type of steroid – known as the Select Androgen Receptor Modulators (SARMS) – which minimizes the undesirable side effects of traditional testosterone on the prostate, liver, hair, and skin by promoting full anabolic activity in muscle and bone, without affecting the reproductive system (Negro-Vilar, 1999; Miner et al., 2007). Many fear that such a medicine would finally concretize the Captain America dream by transforming every Steve Rogers into a super soldier. It has also been reported that DARPA has experimented with medicine that would allow soldiers to feed from their body fat, and that the army has been interested in developing genetically modified food to fight soldiers' fatigue and improve their alertness (Egudo, 2004). The veracity of these allegations has been fuelled by DARPA's almost endless government subsidies and by statements made by the DARPA programme manager Michael Callahan in 2007, to the effect that the research he was supervising aimed to 'bring to battle the same sort of capabilities that nature has given certain animals'; it could be achieved with products such as 'drugs that will boost muscles and energy by a factor of 10, akin to steroids … on steroids' (Lin et al., 2013, pp. 6–7). Stories (Lavine, 2012) and scientific studies (Brunet et al., 2011) have also emerged regarding the use of a new medication called propranolol, which could erase the effects of terrifying memories and treat soldiers who are suffering from post-traumatic stress disorder (PTSD).

As the reader will realize, capacity-restoring and capacity-increasing technologies are very different in their respective aims. While the former are inherently associated with the logic of therapy (which refers to the idea of 'restoring to normal'), enhancement refers to the idea of 'going beyond the normal' (*Beyond Therapy*, 2003, p. 16). The two are also significantly different when viewed from a moral perspective. Indeed, in contrast to capacity-increasing technologies, the use of capacity-restoring technologies or medicines lies on solid ethical principles associated with Modernity. This is what the next section will examine.

Some important moral questions

It is important to emphasize that not only is the recourse to capacity-restoring technologies always ethically acceptable (*Beyond Therapy*, 2003, p. 13), but they can also be viewed as morally necessary and a matter of equality in the realm of political liberalism. One of the great legacies of the Enlightenment revolution of the eighteenth century is undoubtedly our capacity as human beings to pursue our own conception of happiness as a way to express our uniqueness and our authenticity. In doing so, individuals now have the right to follow their own original paths and to realize their goals of self-fulfilment and self-realization: a right that should be equal for everyone no matter what their gender, sexual orientation, or social class. As a consequence, modern liberal societies have established sets of rules that are meant to be universal for their citizens. This was a major shift from the ancient societies of the *Ancien Régime*, in which certain individuals were able to enjoy distinctive rights that were solely based on their belonging to certain classes.[3] As a consequence, the capacity to pursue happiness was, of course, unequal in such societies.

However, such a modern notion of what the Canadian philosopher Charles Taylor calls 'a politics of equal dignity' (Taylor, 1994) remains nonetheless very theoretical, as some individuals may be deprived of their capacity to fulfil their conception of happiness for various reasons. For instance, two individuals whose definition of self-realization consists of pursuing advanced academic studies in order to become university academics and do research may not both be able to achieve this goal, despite the fact that they are equally talented and dedicated to this task. This might simply be because one of them comes from a very rich family that can afford to pay significant tuition fees, while the other comes from a very poor family whose members have barely enough to survive. In such circumstances, and in order to ensure that the latter individual is able to follow her or his conception of a happy life, liberal societies are justified in offering this individual a waiver of tuition fees, as well as a full scholarship. This type of dispensation should not be seen as an undue privilege like those that social classes in the societies of the *Ancien Régime* enjoyed. On the contrary, what Taylor calls a 'politics of difference' (Taylor, 1994) is a way to ensure that the theoretically equal capacities that human beings have to fulfil their own conception of the good life have empirical meaning. This is why many liberal societies have government subsidies that allow citizens who do not have sufficient financial resources to meet the cost of court cases and benefit from free legal aid. Once again, such a special programme is not designed to allow certain citizens to enjoy unjust and unequal advantages over their compatriots. It is a way to make sure that the notion that every citizen in a liberal society should be able to enjoy a fair trial is not simply a theoretical idea. One can indeed easily imagine what this right would mean for someone living in poverty who is unable to afford a lawyer. Without knowing the jurisprudence, the way a trial is organized, and how to cross-examine witnesses, it is obvious that her or his right to enjoy a fair trial would not be the same as that of someone who is able to hire the best lawyers in town. Accordingly, the appointment of a duty solicitor for low-income individuals should not be seen as a privilege, but rather as a means to fully allow them to benefit from a fair and equal trial. Such differentiated treatments grow naturally from the politics of equal dignity as necessary conditions to ensure that our individual right to pursue happiness will not be restricted by external causes that are linked to contingencies over which we may not always have control.

The use of capacity-restoring technologies follows the same pattern. Some individuals are born with physical disabilities or are victims of accidents that prevent them from pursuing what constitutes their conception of happiness on an equal footing with non-disabled people. For instance, an individual's desire to pursue a singing career might be crippled by a hearing impairment. Allowing her or him to benefit from a hearing implant that will give her or him the possibility of following this dream like any other human being is therefore morally justified. For this reason Kenneth Ford and Clark Glymour have argued:

> Almost all people use optional enhancements of their abilities, or optional reductions in their disabilities. Reading glasses, hearing aids, and aspirin tablets remedy common disabilities, from myopia to headaches. Computers are, at base, cognitive prostheses. Diet and exercise regimes can enhance our abilities or remediate disabilities. Prosthetic devices enable those who have lost limbs to grasp, walk, and run. Society finds no moral problem with any of these enhancements. Computerized implants in animal brains have been studied for several decades and have been used in humans without serious ethical qualms for cases of blindness and loss of motor control. There is a degree of cyborgism that is morally tolerated because it helps to compensate for a disability. (Ford and Glymour, 2014, p. 44)

In the military, these forms of technology permit the restoration of lost functionality through the use of prosthetics for soldiers who have been wounded and have lost limbs, and the use of medicines o help them overcome the psychological trauma they have experienced on the battlefield; they act as a way to help individuals to resume normal life when they are discharged from active service.

Of course, this division between therapy and capacity-increasing techniques is not perfect. One of the main problems is undoubtedly the fact that some capacity-restoring technologies may result in allowing the beneficiary to gain supernatural advantages over others. This was notoriously alleged by certain individuals in the case of the former South African runner Oscar Pistorius, who, after being born without fibulas, had both his legs amputated below the knee when he was eleven months old.[4] However, he did not let his disability stop him from practising sports at a high level, and, with the help of artificial J-shaped carbon fibre prosthetic legs, he was able to compete against runners without disabilities. He was nicknamed 'Blade Runner' by the press as his artificial legs appeared to come directly from science fiction or Hollywood movies. As the reader will realize in the coming pages, they many similarities with capacity-increasing technologies that are currently under development or are being used by many armies.

The major difference between Oscar Pistorius's prosthetic legs and the technologies associated with super soldiers lies in the fact that the former are linked with capacity-restoring technologies. They were not originally meant to allow him to benefit from supernatural power as a runner, but rather to let him enjoy a similar life and the same opportunities as athletes without disabilities. In itself, and drawing from the Canadian philosopher Charles Taylor's conception of equality and differences, this desire to compensate someone's incapacity in order to fulfil his conception of happiness on an equal basis with other individuals can only be hailed as something morally just in our modern world.

However, after it was claimed in 2007 by German scientists that Pistorius was using 25 per cent less energy than natural athletes and was therefore benefiting from an unfair advantage over his non-disabled competitors, the South African runner was banned from competing outside Paralympic events by the International Association of Athletics Federations, until this decision was overruled by the Court of Arbitration for Sport, which allowed him to compete at the 2012 Summer

Olympic Games.[5] This decision was followed by controversy among scientists, who disagreed over whether Pistorius indeed enjoyed an undue advantage. Those who were in favour of banning him from competing with 'natural' athletes argued that his limb-repositioning time (the average time a runner takes to move her or his legs from back to front) was much lower (0.28 second, instead of an average of 0.34 second for natural athletes), as was the weight of his lower legs (only 2.4 kg, instead of an average of 5.7 kg) (Eveleth, 2012). They also argued that unlike natural athletes, Pistorius did not suffer from lactic acid in his legs, and that this allowed him to benefit from an advantage in the second part of 400 metre races, where this natural body reaction makes other athletes struggle. The same argument was used by certain experts in order to prevent another amputee athlete, the long-jump athlete Markus Rehm, from fulfilling his dream of participating in the 2016 Rio de Janeiro Olympic Games (BBC, 2016).

Of course, such cases raise important debates that should not be ignored and highlight how moral intentions can become problematic, as the line between capacity-restoring and capacity-increasing technologies can be very thin and blurred. In view of current technological progress, we can predict that international sports governing bodies will have to tackle these issues seriously in the near future. However, this question oversteps the goal of this book. Yet the possibility that capacity-restoring technologies might transform themselves into capacity-increasing technologies has the advantage of illustrating very clearly the prevailing problem with enhancement technologies. Capacity-increasing technologies cannot be justified with the same logic as capacity-restoring technologies, as they tend to provide unnatural advantages to those who are benefiting from them. In this sense, they cannot be envisaged in the light of Taylor's notion of differentiated treatment for the sake of achieving equity, but rather should be seen as a mere privilege – an idea which contradicts one of the moral foundations of our societies. How is this tension transposed into the world of the military? This is what the next section will explore.

Super soldiers and the principle of riskless warfare

When considering the inherent imbalance and inequality that the development and use of capacity-increasing technologies can create between those who can enjoy them and the 'have-nots', certain authors have argued that this would be solid grounds for rejecting them. For instance, Robert Simpson goes as far as to argue that the current trend in the development of super soldiers is morally inadmissible on this account, as it would question the whole logic of permitted killing during warfare (Simpson, 2015). Such a fear is not without merit from a theoretical perspective and, consequently, deserves to be analysed. According to Michael Walzer's account of Just War Theory, the right to kill an enemy combatant relies on a specific account of the right to self-defence. In the civilian world, individuals are allowed to resort to means of action that are generally not permitted in a free and democratic society only if three conditions are met. More specifically, an individual may attack another person's physical integrity only if the former is facing a threat to her or his life; if the aggression is in itself forbidden; and if the danger is real. If these conditions are satisfied, the individual has the right to commit unlawful actions only if there is no other way to escape the danger, if the reaction against the attack is directly posed in order to stop the threat, and if the response is proportionate. If these criteria are met, individuals will not be considered criminally responsible for their actions.

This definition of legitimate self-defence has been extended to a collective definition within international public law and is integral to the right of states to pre-emptively defend themselves against foreign invasion. According to the principles of Just War Theory that surround the morality of starting a war (the principles of *jus ad bellum*), states are justified in defending their sovereignty not only after an attack has taken place against their territorial integrity, but also before such an attack has ever started. This defence will, however, be allowed only if the state has incontrovertible evidence that the threat of invasion is imminent and sufficient. According to Walzer, such a threat has to encompass 'a manifest intent to injure, a degree of active preparation that makes that intent a positive danger, and a general situation in which waiting, or doing anything other than fighting, greatly magnifies the risk' (Walzer, 2006, p. 81).

This notion of pre-emptive self-defence should never be equated with the idea of 'preventive attack', which is a type of warfare which aims solely to deprive an enemy from acquiring, in the long run, the capacity to attack. Such interventions are considered to be illegal under the framework of international law. For the sake of understanding, it is generally admitted that the Six-Day War, started in 1967 by Israel against its neighbours, was a pre-emptive attack, since the Jewish state had strong evidence that Egypt, Syria, and Jordan were about to invade its territory (Walzer, 2006, pp. 82–85). On the other hand, the invasion of Iraq by the United States and its allies in 2003 is generally considered to have been a preventive attack, as Saddam Hussein's regime clearly did not pose an imminent threat to the international community.

In the military world, the criteria are very different. While Walzer admits that only a 'few soldiers are wholeheartedly committed to the business of fighting' (Walzer, 2006, p. 138), this does not change the fact that any member of the armed forces is a legitimate target during wartime and is subject to being killed at any time, notwithstanding her or his willingness to attempt to take an enemy soldier's life. For the American philosopher, the possibility of killing an enemy combatant without being seen as a murderer lies in the fact that combatants belong to a specific class. As he writes:

> Soldiers as a class are set apart from the world of peaceful activity; they are trained to fight, provided with weapons, required to fight on command. No doubt, they do not always fight; nor is war their personal enterprise. But it is the enterprise of their class, and this fact radically distinguishes the individual soldier from the civilians he leaves behind. ... He can be personally attacked only because he already is a fighter. He has been made into a dangerous man, and though his options may have been few, it is nonetheless accurate to say that he has allowed himself to be made into a dangerous man. For that reason, he finds himself endangered. (Walzer, 2006, pp. 144–145)

As members of this class, soldiers from both sides are therefore subject to the risk of being harmed by the enemy. Thus as a soldier, knowing that you can be targeted by your foe at any time, you are entitled to defend yourself by attempting to take her or his life before he or she takes yours. Neither the enemy's moral guilt as a combatant nor the moral reasons for which he or she contributes to his country's motives are relevant. In the case of an unjust war of aggression, only those who are responsible for the beginning of the conflict will have to pay the price of retribution at the end of the war. Combatants are constrained to fight by superior forces (whether political or military) over which they have no control. This is why German, Italian, and Japanese soldiers were not considered war criminals in 1945 for participating in an unjust conflict. Only those who planned crimes against peace were held on trial and sentenced at the Nuremberg and Tokyo trials in the aftermath of the war. By Walzer's account, German soldiers who killed Allied soldiers on the battlefield were as innocent as Allied soldiers who killed German combatants, since they were engaged in a condition of reciprocal killing.

This unique status of soldiers also opens the door to the targeting of those who have been coined 'naked soldiers', for example enemy soldiers who engage in activity that is not battlefield activity and who do not pose a direct threat to the life of anyone (Deakin, 2014, p. 321). Some authors, such as Larry May, have argued that such combatants have regained their legal immunity from being killed (May, 2012, p. 111). Although Walzer admits that shooting an enemy combatant who is taking a bath in a river behind enemy lines might be psychologically troublesome for many individuals, who could come to believe that because of her or his vulnerable condition the soldier in question is no longer an enemy but rather a normal individual, he still believes that this person is a legitimate target. For Walzer, the logic behind such a conclusion lies in the fact that the naked soldier is not similar to an enemy combatant who has surrendered or who is wounded and unable to continue fighting; such soldiers are regaining their status as non-combatants and should, accordingly, be treated with humanity and respect. Not to do so would be a war crime. However, as von Clausewitz has argued, all soldiers – even the naked ones – have abandoned their status as normal individuals by agreeing to join the ranks of the military and by submitting themselves to its martial virtue (von Clausewitz, 1976, p. 144). According to this logic, the naked soldier still remains in a position to harm you or your comrades in the future. It is in this sense that Stephen Deakin, who discusses this idea, writes that 'like a tank, artillery piece, or a military aeroplane, the naked soldier is a weapon of war. Destroying tanks, guns and the like is a legitimate and desirable activity in war whether they are in use or not at the time, and the same is true of naked soldiers' (Deakin, 2014, p. 329).

This traditional definition of self-defence in wartime is pivotal to our understanding of who could legitimately be killed. When this dynamic relationship between combatants is erased, for example when the reciprocal imposition of lethal risks disappears, the whole legitimacy of attempting to take somebody else's life is questioned. This situation has been described by Paul W. Kahn as the 'paradigm of riskless warfare' (Kahn, 2002). According to him, and following Walzer's account of legitimate killing in wartime, there is no moral problem when combatants are in a relationship of mutual risk. However, the issue becomes more problematic when an army is able to destroy its enemies without any risk to its members' lives (Kahn, 2002, p. 3). While the fundamental structure of a war in which soldiers face reciprocal risks to their lives is analogous to a duel where danger is a reciprocal reality (von Clausewitz, 1976, p. 13), the paradigm of riskless warfare has more to do with hunting. In such a situation, the enemy is no more than a prey that can only hide or run: it has no way to escape its gruesome fate.

Such a turnaround in the conceptualization of warfare is intimately linked with the use of drones and other types of uninhabited aerial vehicles (Chamayou, 2015). With such technologies, more advanced armed forces are able to keep a close eye on their enemies and to annihilate them with a single push of a button by an operator located thousands of miles away in a secure environment. For Kahn, the lack of reciprocity engendered by the drones is precisely what makes their use ethically inadmissible.

In his evaluation of capacity-increasing technologies, Simpson draws on the same logic, arguing that combatants benefiting from them will eventually experience a situation where they enjoy a relatively unthreatened position 'which will render them highly resistant to a wide spectrum of normally lethal physical threats, including projectile ammunition, shockwaves, incendiary agents, neurotoxic agents and vesicant agents' (Simpson, 2015, p. 88). However, such a parallel between drones and capacity-increasing technologies is flawed. Of course, these technologies provide their beneficiaries with an asymmetrical advantage over their enemies, but they do not prevent them from being killed. This is what Simpson concedes in his analysis when he states that it will not be 'completely impossible for [the enhanced fighter] to be injured or killed by enemy combatants, but the threat that he poses to the enemy's life drastically outstrips the threat that the enemy poses to his life' (Simpson, 2015, p. 89). However, it is a mistake to confuse an asymmetrical relationship between foes with riskless warfare. These two situations are, of course, far from similar. As mentioned previously, in the case of riskless warfare some combatants are in a position to harm or kill their foes without any similar reciprocal danger to their own safety and lives. In this case, the logic of Walzer's self-defence is jeopardized. On the other hand, in a situation of asymmetrical warfare, despite the fact that some soldiers face less risk to their lives than others, they remain nonetheless vulnerable.

To argue that the quest for asymmetrical warfare is problematic would be to question the morality of almost all previous wars faced by humanity. Indeed, states have historically tried to gain advantages over their enemies. However, this asymmetry has never altered the profound nature of combat: reciprocal danger for the combatants. Moreover, arguing that such a form of relationship is immoral because it reduces soldiers' chances of being harmed on the battlefield would be strictly theoretical and would not consider the subtleties of warfare. More precisely, various historical examples allow us to acknowledge that having an advantage over the enemy is far from a guarantee of victory. A large number of states that have had superior forces on paper or technologies superior to those of their enemy have still been defeated, because military superiority is simply one of the numerous factors that contribute to final victory. It is far from being the determining factor and, depending on the circumstances, it may actually be useless. As examples, we may mention the war between the United States and Vietnam, that between the Soviet Union and Afghanistan, the Battle of Crécy in 1346, during which the heavily armoured French cavalry was destroyed by the outnumbered English archers, or Alexander the Great, who fought King Darius III in 331 BC at the Battle of Gaugamela. In the latter case, despite having an overwhelming force over Alexander, the Persian troops nonetheless lost about fifty thousand men against only four thousand lost by the Macedonian and Hellenic League fighters. The same can be said with regard to the Battle of France in the spring of 1940. Despite having more divisions, more guns, and more tanks, France and its allies nonetheless suffered more than twice as many casualties than Germany. These examples show that asymmetrical warfare is far from similar to riskless warfare. Mere numbers or superior technologies are not in themselves sufficient to guarantee a quick and efficient victory against a foe. Victory is very often connected with more effective strategies.

In addition, we need to understand that war is not similar to sport, where opponents face their opponents on an equal footing. In these latter forms of confrontation, individuals have to obey certain rules in order to prevent players from having unfair advantages over others. This is, for instance, the case in hockey, where the stick can be curved only to a certain degree, in cycling, where the bicycle used cannot have a weight lower than 6.8 kg, or in baseball, where cork bats are forbidden because they increase bat speed and hitting power. In the same vein, sports governing bodies are also fighting (some with more success than others) against doping, which is, of course, another way for athletes to enjoy advantages over others. Why are the worlds of sports and warfare so different in this regard? Why is sport organized as a confrontation in which opponents need to observe the level playing field, and why is that not the case for states fighting one another in warfare?

As will be discussed more at length in the final chapter, the spirit of sport centres on the celebration of the human spirit, body, and mind and is characterized by certain values, such as an ideal of fair play, as ways to experience a healthy life, fun and joy, dedication and commitment, and to transgress one's physical limit. More fundamentally, we can argue that the ethics of sport are animated by a combination of equality and superiority. Although there is an inevitable genetic lottery among individuals, it is assumed that everyone, through training and dedication, will nonetheless be able to overcome this natural physical inequality and be recognized as the strongest, the fastest, or the most skilled individual. These values are the core aspects of this naturalistic Athenian vision of sport. Mere genetic advantages are not sufficient to achieve this goal, and the history of sport is replete with cases of naturally gifted athletes who failed to achieve success simply because of their lack of will. There is an undeniable link between effort and success. This idea that we are all equally able to exceed the talents that nature gave us is a driving force of human beings' endeavour to individually improve their human condition. We all know that we can always push our own limits and that, despite our genetics and natural talents, our capacity to improve is not a mere fantasy. On the contrary, performance-enhancing substances or other forms of cheating that transform any donkey into a fast racehorse are simply shortcuts that alter this ideal. The knowledge that you, as an individual, are able to rise above the rest of the field must derive from the idea that such a possibility is a natural feature that is the same for everyone. The Olympics and other sporting events are celebrations of this ideal: that physical superiority in a specific domain is achievable by everyone and depends only on their willingness to train hard and be dedicated in reaching their objective. It is this ideal that makes achieving one's goals through artificial and non-natural means a form of partial alienation from our own self-conscious and self-directed doings. In this sense, dopers are becoming partly estranged from humanity.

The ethical dimension is not the same in cases of armed conflict between nations, simply because the modern Machiavellian world of politics is amoral. It is, rather, dominated by an ideal of pure pragmatism, which allows political leaders to choose all possible options in order to keep domestic politics peaceful and well ordered, while increasing or maintaining their status in the world of international politics. All the means used by political leaders are mere instruments that serve these higher purposes. This is why von Clausewitz's famous remark about war being the continuation of policy by other means is so significant (von Clausewitz, 1976, p. 28). More specifically, the interactions between nations are not dominated by the ideal of equality inherent to sport. On the contrary, states feel that they are entitled to increase their power by cheating other states, whether through spying or similar means. Moreover, some states, because of asymmetries of power, are able to coerce other nations to change some of their policies. In this perspective, military power is simply a corollary of politics. Therefore, capacity-increasing technologies, such as those used by aviation, infantry, or naval forces, are actually a constitutive part of modern politics. Given the amoral nature of this world, enjoying advantages in order to maintain or increase state power is not in itself ethically problematic. That will be the case only if a state that benefits from such a military advantage uses it according to the international laws that apply in international politics and that they have voluntarily agreed to respect.

When we think of such technologies, it seems clear that we are more in the sphere of asymmetrical advantages than of riskless warfare. It is obvious that exoskeletons and medicines currently being used provide soldiers with an edge over their opponent who may not have the possibility to enjoy them, but they do not make them invulnerable to death, as conceded by Simpson himself in the aforementioned remark. If we were to conclude otherwise, we would have to prevent soldiers from using bulletproof vests, armoured vehicles when operating in dangerous zones filled with improvised explosive devices (IEDs), or night vision lenses that allow soldiers to see the enemy in the dark without being seen. Capacity-increasing technologies currently being developed by the various armed forces are similar to these latter forms of armaments. While they may increase their chances of survival on a battlefield, they do not create a situation where the reciprocity of death will be eliminated. For the reasons previously evoked, it would then be a mistake to automatically equate asymmetrical military advantages with something that should be considered unethical. It is in this perspective that we can agree with Hillary F. Jaeger, who wrote that 'one does not normally think of performance enhancement as cheating in a military operational context; rather, the search for asymmetric advantages, within the bounds of the Law of War, is both good strategy and sound tactics' (Jaeger, 2007, p. B129). As long as the technology being used does not run counter to international conventions (such as those on the use of poisonous gas and bacteriological methods of warfare, or ways of starting an unjustified war of aggression), military capacity-increasing technologies for combatants are not *prima facie* unethical.

However, even though capacity-increasing technologies can be morally defended on the basis of the military institution's duty of care and the fact that they do not – unlike drones – contribute to making the soldiers who are using them absolutely invulnerable, this fact has nonetheless led authors to argue that this situation might be harmful to the principles of Just War Theory. More precisely, it has been argued that armies benefiting from them will probably see a decreased rate of casualties within their ranks (Beard, Galliott, and Lynch, 2016). While this situation has significant positive moral implications, we must remember that it might lead some countries to see the recourse to war – which should, according to Just War Theory, always be a last-resort option – as a low-cost decision from a political perspective. Indeed, by having the capacity to deploy soldiers with a decrease chanced of being harmed or killed, some states might come to believe that it is easier to resolve their conflicts with other nations through military means rather than diplomacy. In this sense, we must be aware that capacity-increasing technologies, while serving a higher moral end, may nonetheless contribute to harming the principles of Just War Theory by lowering the threshold for war. Although this fear cannot be ignored, we should not exaggerate it, because there is no evidence to back it up (Leveringhaus, 2016, pp. 13–14). As mentioned previously, the use of a superior armament does not necessarily mean an assured victory over a weaker foe. History provides us with numerous examples in this regard.

Conclusion

Despite the fact that the moral reasons for using capacity-restoring technologies are very different from those for using capacity-increasing technologies, it would nonetheless be a mistake to automatically condemn the latter. Of course, they create an asymmetrical relationship between the states that have them and those that do not. However, as argued in this chapter, this inequality between combatants does not challenge the moral justification for killing during wartime. Indeed, capacity-increasing technologies allow some soldiers to augment their chances of survival on the battlefield, but they do not create a situation where their beneficiaries will become absolutely invulnerable to being wounded or killed. In terms of the morality of warfare, this distinction, which is not understood by everyone in the literature, is the central element that allows individuals to kill other human beings in times of war. Therefore, as long as these technologies do not engender invulnerability, they should not be considered to be morally reprehensible. Of course, if they were ever to create situations in which their beneficiaries became absolutely immune from being wounded or killed, the whole question would have to be debated again. However, the current trend in the development of these technologies does not show that it is currently the case.

Moreover, as the next chapter will argue, the fact that these technologies may increase soldiers' chances of survival on the battlefield can further justify their recourse, as they can help the military institution to satisfy one of its moral obligations, namely, its often neglected duty to care for and ensure maximal protection for the health and life of its service personnel. This obligation is another element that militates in favour of developing and using capacity-increasing technologies in the military and another reason why they should not be dismissed as being morally wrong.

Notes

1 This confusion between enhancement and therapy is unfortunately present in the analysis of some authors (Hutchinson and Rogers, 2015).

2 For instance, even before Pervitin was used by the German military during the war, it was already widely used by German civilians. This was made possible by the commercial strategy of its producer, the Temmler firm, which presented it as a way to help fight the stress of modern life, to increase women's sex drive, and to lose weight. As reported by Norman Ohler, 'Whether it was secretaries who wanted to type faster, actors who wanted to increase their focus before a play, writers who wished to sharpen their mind or workers who wanted to increase their production, Pervitine quickly spread through all social classes prior to the war' (Ohler, 2016, p. 44).

3 For instance, before the revolution of 1789, French nobles were granted an exemption from paying certain taxes; some ecclesiastic, civic, and military positions were reserved for them, and they also enjoyed a number of exclusive privileges.

4 The possibility of capacity-restoring technologies providing above-average capacities to its bearers is not restricted to the sport domain. For instance, Aimée Mullins, who suffers from the same disability as Pistorius, had a successful career as a model with the help of her artificial legs. Indeed, in contrast to models without disabilities, she is able to modify at will the appearance of her legs as well as her size. This is also the case for the British artist Neil Harbisson, who is also referred as the world's first cyborg artist. Born with an extreme form of colour blindness, he had a special antenna implanted into his skull in 2004 that now allows him to perceive colours through vibrations.

5 Contrary to what many people think, Pistorius was not the first athlete with a disability to compete against non-disabled athletes. We might also mention the one-legged gymnast George Eyser, who competed in the 1904 Olympic Games, Marla Runyan, a visually impaired runner who competed in the 2000 Olympics, or Natalia Partyka, an amputee tennis player who competed in the 2008 and 2012 Olympics.

References

BBC. 2016. 'Rio Olympics: Paralympian Markus Rehm will not Compete in Rio Long Jump', 2 July. www.bbc.com/sport/olympics/36565093 (last accessed 27 September 2017).

Beard, Matthew, Jai Galliott, and Sandra Lynch. 2016. 'Soldier Enhancement: Ethical Risks and Opportunities', *Australian Army Journal*, Vol. 13, No. 1, pp. 5–20.

Beyond Therapy: Biotechnology and the Pursuit of Happiness. 2003. A Report of the President's Council on Bioethics. Washington, DC: Dana Press.

Brunet, Alain et al. 2011. 'Trauma Reactivation under the Influence of Propranolol Decreases Posttraumatic Stress Symptoms and Disorder', *Journal of Clinical Psychopharmacology*, Vol. 31, pp. 547–550.

Chamayou, Grégoire. 2015. *A Theory of the Drone.* New York: New Press.

Deakin, Stephen. 2014. 'Naked Soldiers and the Principle of Discrimination', *Journal of Military Ethics*, Vol. 13, No. 4, pp. 320–330.

Egudo, Margaret. 2004. *Overview of Biotechnology Futures: Possible Applications to Land Force Development.* Canberra: Australian Government Department of Defence.

Ehrenberg, Alain. 1994. *Le culte de la performance.* Paris: Calmann-Lévy.

Eveleth, Rose. 2012. 'Should Oscar Pistorius's Prosthetic Legs Disqualify him from the Olympics?', *Scientific American*, 24 July.

Ford, Kenneth and Clark Glymour. 2014. 'The Enhanced Warfighter', *Bulletin of Atomic Scientists*, Vol. 70, No. 1, pp. 43–53.

Higgins, Chris. 2014. 'Darpa Jetpack to Help Soldiers Run Four-Minute Miles', *Wired*, 17 September.

Hutchinson, Katrina and Wendy Rogers. 2015. 'Ethical Considerations in Military Surgical Innovation'. In Jai Galliott and Mianna Lotz (eds), *Super Soldiers: The Ethical, Legal and Social Implications.* Farnham, UK: Ashgate, pp. 121–140.

Jaeger, Hillary F. 2007. 'A Glance at the Tip of the Iceberg: Commentary on Recommendations for the Ethical use of Pharmaceutical Fatigue Counter-Measures in the US Military', *Aviation, Space, and Environmental Medicine*, 78 (5, sec. II), pp. B128–B130.

Kahn, Paul W. 2002. 'The Paradox of Riskless Warfare', *Philosophy & Public Policy Quarterly*, Vol. 22, No. 3 (Summer), pp. 2–7.

Lavine, Robert. 2012. 'Ending the Nightmares: How Drug Treatment could Finally Stop PTSD', *The Atlantic*, 1 February.

Leveringhaus, Alex. 2016. *Ethics and Autonomous Weapons.* Oxford: Palgrave Macmillan.

Lin, Patrick. 2010. 'Ethical Blowback from Emerging Technologies', *Journal of Military Ethics*, Vol. 9, No. 4, pp. 313–331.

Lin, Patrick, Maxwell J. Mehlman, and Keith Abney. 2013. *Enhanced Warfighters: Risk, Ethics, and Policy.* The Greenwall Foundation. http://ethics.calpoly.edu/Greenwall_report.pdf (last accessed 8 October 2017).

May, Larry. 2012. *After War Ends: A Philosophical Perspective.* Cambridge: Cambridge University Press.

Miner, J. N. et al. 2007. 'An Orally Active Selective Androgen Receptor Modulator is Efficacious on Bone, Muscle, and Sex Function with Reduced Impact on Prostate', *Endocrinology*, Vol. 148, pp. 363–373.

Negro-Vilar, Andres. 1999. 'Selective Androgen Receptor Modulators (SARMs): A Novel Approach to Androgen Therapy for the New Millennium', *Journal of Clinical Endocrinology and Metabolism*, Vol. 84, pp. 3459–3462.

Ohler, Norman. 2016. *L'extase totale: le IIIe Reich, les Allemands et la drogue.* Paris: La Découverte.

Rupprecht, Rainer et al. 2009. 'Translocator Protein (18 kD) as Target for Anxiolytics Without Benzodiazepine-Like Side Effects', *Science*, Vol. 325, pp. 490–493.

Simpson, Robert Mark. 2015. 'Super Soldiers and Technological Asymmetry'. In Jai Galliott and Mianna Lotz (eds), *Super Soldiers: The Ethical, Legal and Social Implications*. Farnham, UK: Ashgate, pp. 81–91.

Taylor, Charles. 1994. 'The Politics of Recognition'. In Amy Gutmann (ed.), *Multiculturalism*. Princeton: Princeton University Press, pp. 25–73.

Tyler, William J. 2010. 'Remote Control of Brain Activity Using Ultrasound'. http:// science.dodlive.mil/2010/09/01/remote-control-of-brain-activity-using-ultrasound/ (last accessed 27 September 2017).

Von Clausewitz, Carl. 1976. *On War*. Oxford: Oxford University Press.

Walzer, Michael. 2006. *Just and Unjust Wars. A Moral Argument with Historical Illustrations*. 4th edition. New York: Basic Books.

2

The moral obligation of soldier enhancement research

The muster-roll of the dead may be a monument of governmental incapacity as well as a certificate of patriotism and courage. It is always glorious for the other man to die for his country, – at least the survivor says so; but the fact that his life has been needlessly thrown away is calculated to throw some doubt on the subject. A civilized nation cannot afford to throw away a single life.

Captain James Chester, 1883

For many people, the prospect of super soldiers can be quite frightening. The current developments in military technology seem to indicate that we are about to experience a paradigm shift in warfare, namely that soldiers will look more like robots with supernatural features than like conventional human beings as in the past. This perspective has been summed up by Ryan Tonkens, who wrote the following about super soldiers:

That we might need to change anything substantial about human beings so that they can be (more) effective in their militarist roles is deeply unsettling. It implies that war is becoming so complicated, rapid, and foggy that human soldiers in their unaltered state cannot adequately keep up with its pace and demands. Perhaps this has always been true, although the situation seems exacerbated in the context of contemporary warfare, accompanied by ever-evolving technological capabilities. (Tonkens, 2015, p. 53)

As this book will show in the coming chapters, this fear generates many significant moral questions that need to be addressed before societies fully embark on this new path. One of them is linked with the fact that capacity-increasing technologies will allow some soldiers to benefit from advantages over their foes who do not benefit from them. However, as has been argued already, it would be a mistake to conclude automatically that they should not be used, as long as they are allowing their beneficiaries only to limit their vulnerability and not to eliminate it completely.

This chapter will now consider the potentialities of capacity-increasing technologies as necessary moral goods. First, allowing soldiers to benefit from capacity-increasing technologies can be seen as a moral duty of the military, which, like other types of employers, has the obligation to limit as far as possible the dangers to its soldiers' lives and health. Second, the international rules of warfare also stipulate that soldiers have to refrain from committing certain actions on the battlefield, such as deliberately harming civilians and individuals who are no longer considered to be combatants. Unfortunately, despite the best efforts by the military to train its members to act in accordance with these rules, the psychological consequences that soldiers very often encounter on the battlefield create situations that make them irresponsive to these moral norms. As this chapter will show, the My Lai massacre during the Vietnam War and more recently the war in Iraq are good examples of this. Therefore we also cannot ignore the prospect that capacity-increasing technologies might prevent the appearance of these psychological problems and contribute positively to respect for these moral rules of warfare. These two reasons contribute to strengthening the need to develop and use capacity-increasing technologies in the military from a moral perspective.

The moral obligation to use capacity-increasing technologies

Given the nature of what seems to be occurring in the military, it is normal that a large number of people should have important concerns regarding the use of capacity-increasing technologies. A simple search on the Internet provides numerous blogs and websites dedicated to conspiracy theories. Why is this so? It is largely because history is full of awful stories of soldiers who have been used as guinea pigs by the military, which have been brought to light in the form of lawsuits and congressional investigations. These stories tend to show soldiers being treated as expendable in the name of science. We might mention the various atmospheric nuclear tests conducted by the United States – such as Operations Plumbbob, Castle Bravo, or Crossroads – during which thousands of soldiers were deliberately exposed to radiation. The Soviets did the same in 1954 during the Totskoye nuclear exercise, in which forty-five thousand soldiers were exposed to an atomic explosion twice as powerful as the one at Hiroshima.

The Edgewood Arsenal experiments are also famous for having tested various chemical substances on more than five thousand soldiers over a period of twenty years. [1] Psychotropic and hallucinogenic drugs such as PCP, LSD, and 3-quinuclidinyl benzilate – commonly known as BZ – were tested with the aim of creating operational weapons that could incapacitate enemy soldiers (Ketchum, 2012; Khatchadourian, 2012). It is reported that the Soviets were performing clinical tests with similar drugs. From these examples, it is clear that the military has not hesitated to overrule ethical norms with regard to this type of research. This is why many feel that the military-industrial complex is untrustworthy when it comes to the treatment of soldiers as individuals, when according to Kant's moral philosophy they should never be treated merely as a means, but rather as an end in themselves.

These famous cases illustrate that it has always been the goal of the military-industrial complex to enjoy military superiority over its potential foes, and that this desire has always been pursued without any consideration for soldiers' lives or health. The same can be said regarding the use of Pervitin by the Germans during the Second World War, mentioned above. In the name of military effectiveness, the state saw the numerous negative long-term effects of the pill as a necessary evil, even though some soldiers became addicted to the drug and suffered dizziness, depression, and hallucinations, while others died of heart failure or shot themselves or their comrades during psychotic phases (Hurst, 2013).

These historical examples should not be taken lightly and should not lead us to neglect the ethical reasons that should justify their use. The need is then to determine when these technologies can be legitimately used from a moral perspective. One way to justify them lies in the dynamic relationship between employers and employees: more specifically, in the duty on the part of the former to care for the latter. Indeed, in professional organizations, employees have to voluntarily fulfil numerous responsibilities towards the company or institution to which they are attached. For instance, employees owe to the employer the duty to be honest, to carry out orders given by the employer (as long as they are legal), not to disclose confidential information, to work with care and professionalism, and to look after the employer's property. In return, employers owe duties to their employees, such as the obligation to pay them for their work, to give them the correct information about the rights and privileges in their work contracts, and to allow reasonable opportunity to ensure that their complaints are considered. However, the most important duty owed by employers is undoubtedly their obligation to do everything in their power to create safe working conditions. In certain jurisdictions, employees even have the right to refuse to fulfil their duties if they believe that they might otherwise lead to risks to their health or life. Of course, what constitutes an unacceptable risk or danger is a very contentious notion, but it is normally considered to be an imminent danger to the health and safety of the worker carrying out the occupation or for other workers present at the work site. Organizations, including governments, can be held responsible for negligence if they act in a way that threatens the safety of their employees.

Naturally, this duty of care varies, depending on the nature of the work. For instance, the right to refuse to work is limited for individuals whose profession is inherently associated with a dangerous mandate, such as fire fighters or police officers. However, it does not mean that individuals carrying out these dangerous functions can be treated by their employers as expendable goods. While it is true that they voluntarily accept exposing themselves to risks to their lives, nonetheless they remain individuals whose right to life should be guaranteed with all possible means. For instance, in the United States the Massachusetts legislature has adopted a state-funded Bulletproof Vest Reimbursement Program, which allows its police officers to be refunded for purchasing protective vests. This programme is described as the 'state's commitment to providing Massachusetts police officers with the best lifesaving equipment available' ('Bulletproof Vest Program', n.d.). The same logic applies to the state of Illinois, which, in conformity with the Law Enforcement Officer Bulletproof Vest Act, must provide a bulletproof vest for each of its law enforcement officers. Fire fighters are also provided with protective gear in order to help lower their risks while carrying out their duties. For instance, in the United Kingdom, fire and rescue authorities are required to provide sufficient equipment and training for individuals who perform such duties in order to let them fulfil their professional obligations in the safest way possible.

These examples show that despite the fact that certain individuals voluntarily agree to perform life-threatening duties, it does not restrict the requirement for their employers to provide all the equipment and training that will allow them to fulfil their duties with minimal risks to their health and safety. It would indeed seem unacceptable for authorities to ask a police officer to fight crime without the necessary training, a weapon, a bulletproof vest, and a vehicle. In fact, asking someone to perform dangerous duties in such circumstances would be seen as gross negligence.

In this respect there is little difference between fire fighters, police officers, and members of the military. Just like the former groups, soldiers willingly accept making the ultimate sacrifice in the fulfilment of their duties. However, this possibility should remain only a last resort, and it would be highly problematic to treat members of the military as simple cannon fodder. While certain cases in history tend to challenge this idea (for instance the Battle of the Somme in 1916, led by British Field Marshall Douglas Haig,[2] or the Chemin des Dames offensive in 1917, directed by French General Robert Nivelles, which led to deaths and casualties of 190,000 soldiers for minimal gain of ground), it is nonetheless clear that the military as an institution has always tried to provide its members with the necessary protective equipment and to update it with regard to the evolving nature of warfare. This was the case shortly after the beginning of the First World War when the various European armies, engaged in fierce battles on the Western front, realized that their soldiers were being exposed to lethal risks of head wounds due to shrapnel and other artillery shells. They quickly replaced the soldiers' cloth caps with the now famous Adrien, Brodie, and *Stahlhelm* helmets. To have done otherwise would have been synonymous with failure to ensure the best possible protection for soldiers' safety and respect for the value of their lives. With time, this care for soldiers' lives and health has only increased.

However, even though military personnel are considered to be employees in the United States, they are nonetheless excluded from claiming compensation for negligence on the part of the military if they suffer injuries or trauma arising out of combat activities. This rule also bars family members of service personnel killed in action from suing the government for wrongful death or negligence.[3] This judicial notion, which immunizes the military and the government from lawsuits by soldiers who have been negligently injured during military service, finds its roots in the idea that 'the King can do no wrong' and was codified in the 1950 Feres v. United States decision by the US Supreme Court. Such a disposition is fairly easy to understand: knowing that they might be held responsible for a bad decision (even if it initially appeared to be a good one), commanding officers might become fearful of ordering their soldiers to perform dangerous tasks. This sword of Damocles would for all practical purposes freeze the whole chain of command and make the military non-operational.

Despite this judicial norm, it is not the case that the military has no duty of care towards its members. In fact, its combat practices seem to reflect the desire to preserve their lives as far as possible. For instance, soldiers are very often denied their own requests to perform missions that are very likely to lead to death. Good examples of this are the cases of Master Sergeant Gary Gordon and Sergeant First Class Randall Shughart during the Battle of Mogadishu in 1993. After a second Black Hawk helicopter (call sign 'Super Six-Four') was shot down by Somali militias, these two men asked to be placed near the crash site in order to protect the four critically wounded personnel, despite being well aware of the growing number of enemy personnel closing in on the site. Their commanding officer refused their request twice, knowing that their demand was suicidal. However, after their third request they were finally allowed to be placed at the site, which unfortunately led to their deaths. For their courage, they were both awarded the Congressional Medal of Honor, the highest military honour given by the United States for personal acts of valour on the battlefield. This example shows that military commanders believe that they have an implicit obligation not to expose the lives of their soldiers to unnecessary risks and that, if it such risks are inevitable, they have to be the result of a voluntary and informed decision on the part of the soldiers.

This duty of care is now a pivotal component in the military's rhetoric and that of political elites. They fully realize that medical and psychological problems deriving from military missions – such as stress and fatigue – often have deadly consequences even for the best-trained soldiers. As was argued by a DARPA official, 'Sleep, nutrition, activity level … they all increase the possibility of a kid getting killed on the battlefield. We tried to figure out how to make those not be rate-limiting factors … so these kids could perform at their peak, stay at their peak, and come home to their families' (Shachtman, 2007). Of course, the explicit statement by elected officials that 'the well-being of our soldiers must remain the first priority of a state' (Cohen and Shelton, 1999, p. 27) may not give the true reason why professional military organizations are keen not to expose their members' lives to risk unnecessarily. It may be explained by the fact that a significant number of casualties might negatively impact the recruitment of potential candidates, who might back out from joining the military, thinking that the risk to their lives is too great. Similarly, in a world dominated since the Vietnam War by the fear of military losses, political elites are pressurizing members of the military high command to avoid even minimal losses because of the possible negative repercussions on public opinion. It may also be due to the fact that professional military institutionss have spent many years and much money on the development of soldiers and that risking their lives engenders a significant cost in terms of time and resources for their replacement. Indeed, as highlighted by Patrick Lin, Maxwell J. Mehlman, and Keith Abney, 'Some estimates put the United States government's investment in each soldier, not including salary, at approximately $1 million, helping to make the US military the best equipped in the world; nonetheless, that soldier is still largely vulnerable to a fatal wound delivered by a single 25-cent bullet' (Lin, Mehlman, and Abney, 2013, p. 1). Moreover, the desire to create super soldiers can also be explained by another monetary concern. Indeed, with the help of these capacity-increasing technologies, combatants will also be available over longer periods of time, with correspondingly increased capacity to perform dangerous missions. This may allow the military institution to complete more missions while needing a decreased number of soldiers at its disposal.

However, these utilitarian and strategic considerations should always remain secondary to the duty of care, which is now regarded in the literature as a moral obligation on the part of the state and the military institution. Despite the fact that soldiers have a duty to fulfil their tasks for their employer (which can go as far as sacrificing their lives), the military institution, like any other employer, also has an obligation to protect them as much as possible. More precisely, it must ensure that soldiers will benefit from training and equipment that will allow them to carry out what they are ordered to do in a way that will expose them to the least possible risk to their safety and life;[4] this is intimately connected with what Bradley Jay Strawser refers to as the 'principle of unnecessary risk (PUR)'. According to him, the military can legitimately order its members only to carry out tasks that do not violate the principles of a Just War and that do not command them to take unnecessary, potentially lethal risks or any risk of bodily harm (Strawser, 2010, p. 344). Strawser believes that the obligation to care for soldiers' lives as much as possible is uncontroversial. Only the accomplishment of a highly superior goal that could not be gained through less risky means, such as the capture of a terrorist leader whose knowledge of information is considered to be invaluable for deterring future attacks from the group he or she is leading, could justify overriding this principle. The case of Osama bin Laden is a good example of this. The United States had strong reasons to believe that the leader of al-Qaida was hiding in his compound near Abbottabad, Pakistan, and his killing would have been less risky if the US Army had used a drone instead of sending six helicopters and seventy-nine members of the Special Forces and the Central Intelligence Agency.[5] The main reasons invoked by President Obama were the fear of collateral damage,[6] the desire to identify his body and to prove, once and for all, that bin Laden was in fact dead, and the fear that he might not be in the compound (McNeal, 2011). However, according to the president, a major argument in favour of sending in troops on the ground – despite the potential deadly implications – was the sheer intelligence value of the compound. As Admiral Bill McRaven, the mission commander, stated, 'there might be the possibility that we would get enough intelligence out of the compound, even in a very short operation, that would help us dismantle other portions of the organization' (Bowden, 2012). According to Strawser's theory, such a possibility certainly qualifies as a 'highly superior goal' justifying the momentary subordination of the principle of unnecessary risk.

While this principle is regarded by Strawser as a way to justify the use of uninhabited aerial vehicles such as drones, it is possible to extend it to the military's obligation to minimize soldiers' exposure to potentially lethal circumstances. For instance, if an army considers it necessary to send troops on potentially danger-ous missions on narrow roads filled with improvised explosive devices (IEDs) in uncontrolled territories, such as those experienced by the International Security Assistance Force in Afghanistan (ISAF) or in Iraq, it is unconscionable to ask them to fulfil these missions without the necessary equipment. This might include fully armoured vehicles able to sustain the impact of IEDs,[7] land mine detectors, and personal protective devices, as well as sufficient training that allows them to react in the most efficient way if they are victims of such attacks. In view of the known deadly threat of IEDs in counter-insurgency situations, it would be careless for the military to send its troops on such dangerous missions without offering the support of such equipment, and negligent to deploy them without the sufficient protective material at their disposal. It was in this area that the former British government of Tony Blair was severely criticized in the Chilcot Report of July 2016 for providing its troops serving in Iraq with the Snatch Land Rover, which was completely unsuitable for the type of threats they were facing. As stated in the report, the British Ministry of Defence knew before the deployment in Iraq that this Protected Patrol Vehicle afforded only limited protection.[8] According to Lieutenant General Graeme Lamb, the British army did not use the Snatch Land Rover for missions in Northern Ireland when there was a threat of massive IEDs. Despite the fact that a significant and serious threat to British forces serving in Iraq emerged as early as July 2003 and that military officials informed the government clearly that the army did not have wheeled vehicles with sufficient armour against insurgent forces and IEDs, the government failed to replace the Snatch Land Rover with a vehicle providing more protection. The consequences were deadly: thirty-seven personnel of the British army were killed by roadside bombs while using this vehicle, which prompted soldiers to label it the 'mobile coffin'.

This particular example raised the question of a state's responsibility to provide its soldiers with the safest equipment in order to protect their health. The traditional jurisprudence that prevents governments from being held responsible for negligence towards their service personnel seems to be evolving in a direction that is more similar to the duty of care they owe to individuals who are not injured or killed in situations arising out of combatant activities. In fact, families of soldiers who lost their lives on board the Snatch Land Rover successfully filed a lawsuit against the Ministry of Defence, arguing that the department had failed to provide suitable armoured equipment to protect against IEDs and had consequently breached Article 2 of the European Convention on Human Rights (Right to Life).[9] Although the lawyers representing the Ministry of Defence raised the argument that 'the King can do no wrong', the judges rejected this claim and stated that the armed forces owed a duty of care under the law of negligence and explained that the department had failed to properly equip the tanks and to give the soldiers adequate training. From the perspective of historical warfare, this decision, of course, represents a major jurisprudential shift, since we tend (wrongly) to believe in light of certain famous battles – Gettysburg, the Somme, or the Chemin des Dames, just to name a few – that the military can treat its members as sub-humans with no rights simply because they voluntarily agree to risk their lives for their country. On the contrary, the decision simply confirmed a conventional logic that employers owe a duty of care to their employees and an obligation to provide them with the necessary equipment and training in order to limit, as far as possible, their exposure to harm or death.

Many capacity-increasing technologies are specifically destined to fulfil this moral obligation that the military institution has as an employer. This is especially true of the various types of exoskeleton being developed not only by the United States but also by France. One of the main goals of such equipment is to relieve soldiers from the strain of the weight they have to carry on the battlefield. It has been reported that although the US Army Science Board recommends that soldiers carry no more than 50 pounds, soldiers may carry an average load of 87 to 127 pounds while on extended foot patrol (Bernton, 2011). Consequently, the army has observed an increase in the number of veterans who have served in Iraq and Afghanistan retiring with major musculoskeletal injuries, such as degenerative arthritis, cervical strains, connective tissue, or spinal injuries, that very often make it impossible for them to resume normal civilian life without significant physical impairment. The use of exoskeletons could contribute to resolving this important issue. In a sense, the exoskeleton is simply an evolution in the logic of limiting the weight that soldiers have to carry, and in this sense the wearing of this equipment is very similar to the use of mule-drawn carts. In view of the nature and aims of exoskeletons, it is no wonder that other non-military organizations, such as police forces and fire-fighting squads, are also considering using this equipment to protect workers from suffering from muscle strain injuries (Lavallée, 2016; Oakley, 2015). We might also mention a special type of contact lens that is currently under development (Anthony, 2012). Such devices are intended to provide give soldiers access to important information without losing their attention to their very often deadly surroundings. For instance, instead of having to keep their eyes off the battlefield in order to study a detailed map, soldiers using the special contact lenses would be able to see the details of the terrain while retaining a total view of it, just as in a typical first-person shooter video game. Moreover, this technology 'could also potentially provide real-time feeds from satellites, drones, or even a teammate's helmet camera, displayed as if on a translucent computer screen to a fighter on the move' (Barrie, 2012).

The desire to use these forms of capacity-increasing technologies is also intimately connected with the prospect that they might contribute to minimizing soldiers' exposure on the battlefield. This idea plays an integral role in the development of jetpacks that would allow soldiers to run faster. As explained by one of its engineers, 'If you think of a Navy SEAL or an Army soldier that has to get in somewhere quick and do whatever they've gotta do, but maybe get out of there just as quickly, so these devices can really help soldiers to not only accomplish their goals and succeed in their missions, but potentially save human lives as well' (Loria, 2014).

We can also say that the use of medicines and other technologies motivated by the desire to reduce fear among soldiers and to make war a less traumatic experience draws from the same logic. This is clearly the principle behind the use of drugs like XBD173 or propranolol. Their development has the clear intention of treating PTSD, which unfortunately affects thousands of veterans, often making them unable to resume normal life once they retire (Gardner and Griffiths, 2014; Bell, 2008). According to the US Department of Veterans Affairs, between 11 and 20 per cent of American soldiers who served in Operation Iraqi Freedom and Operation Enduring Freedom in Afghanistan have PTSD in a given year, while the proportion is about 12 per cent for those who served in 1991 during Operation Desert Storm. In addition, it is estimated that about 30 per cent of Vietnam veterans have had PTSD in their lifetime.[10] In view of the disastrous and long-term effects of this affliction for the soldiers and their families, such as nightmares, the inability to cope with normal interpersonal social interactions, high rates of suicide, marital problems, and family violence, the use of these medicines can also be seen as an obligation on the part of the military institution. If it is common knowledge that many soldiers deployed in combat zones return home mentally disabled and unable to lead normal lives afterwards, then it can be argued that not acting in order to limit or prevent such a situation, despite having the tools to do so, would be tantamount to a form of negligence and a lack of respect for the soldiers as human beings able to pursue happiness in the way they want to. On the contrary, the military institution owes an extended – although not total – duty of care to its members.[11]

The prospect of morality-increasing technologies

As discussed in the next chapter, while there are valid reasons to fear that capacity-increasing technologies may harm soldiers' moral agency and their duty to disobey illegal or immoral orders, it is also necessary to evaluate the prospect of technologies that might increase soldiers' sense of morality if we wish to have a balanced and objective view on the matter. Even though this obligation on the part of combatants has evolved in a positive direction since the Second World War as a way to eliminate the most repulsive aspect of mankind during warfare and to give more humanity to conflicts, we must face the fact that this goal has not been transformed into reality. On the contrary, atrocities are still far more common than we would like to think.

More precisely, even if we cannot quantify the emotional strain of combat on soldiers (Bourke, 1999, p. 232) or scientifically analyse the degree of ethical behaviour on the battlefield (Office of the Surgeon General, 2006), it is nonetheless possible to estimate it and its consequences on the respect for the laws of war on the basis of data collected by the military. For instance, during the Vietnam War, a study has 'pointed out that for all men in heavy combat, one-third of men in moderate combat, and 8% in light combat had seen atrocities or committed or abetted non-combatant murder' (quoted in Arkin, 2010, p. 335). Moreover, a study conducted by the Office of the Surgeon General in 2006 has also shown a serious lack of ethical behaviour on the part of American soldiers and marines who served during Operation Iraqi Freedom (Office of the Surgeon General, 2006, pp. 35–40). Its main conclusions in this regard deserve to be shown:

- Only 47 per cent of soldiers and only 38 per cent of marines agreed that non-combatants should be treated with dignity and respect;
- Well over a third of soldiers and marines reported that torture should be allowed, whether to save the life of a fellow soldier or marine (41 per cent and 44 per cent, respectively) or to obtain important information about insurgents (36 per cent and 39 per cent);
- The most common unethical behaviour that soldiers and marines reported engaging in was swearing at and/or insulting Iraqi non-combatants in their

presence, with 28 per cent of soldiers and 30 per cent of marines reporting doing this;

- Far fewer soldiers and marines reported damaging or destroying Iraqi property when it was not necessary (9 per cent and 12 per cent respectively), or hitting or kicking a non-combatant when it was not necessary (4 per cent and 7 per cent respectively);
- The battlefield ethics violation that soldiers and marines were most likely to report included a unit member injuring or killing an innocent non-combatant, with 55 per cent of soldiers agreeing that they would report a unit member and 40 per cent of marines agreeing that they would report a fellow marine;
- Soldiers and marines were least likely to report a unit member for unnecessarily destroying or damaging private property, with 43 per cent of soldiers indicating that they would report a unit member and 30 per cent of marines indicating that they would do so;
- Fewer than half of soldiers and marines would report a team member for unethical behaviour, with the marines less likely to report fellow marines than soldiers were to report fellow soldiers;
- A large majority of soldiers and marines reported that they received training in how they should treat non-combatants, yet a third of marines and over a quarter of soldiers did not agree that their non-commissioned officers (NCOs) and officers made it clear that they should not mistreat non-combatants;
- Soldiers and marines were more likely to report engaging in the mistreatment of Iraqi non-combatants when they were angry. For all the behaviours under study, soldiers and marines who had high levels of anger were twice as likely to engage in unethical behaviours on the battlefield as those soldiers and marines who had low levels of anger;
- Soldiers who had a member of their unit become a casualty were more likely than those who did not to engage in insulting or swearing at non-combatants in their presence and to report damaging or destroying Iraqi property when it was not necessary;
- Soldiers who handled dead bodies or human remains were also more likely to have insulted or sworn at non-combatants in their presence, damaged or destroyed Iraq property when it was not necessary, or physically hit or kicked a non-combatant when it was not necessary than soldiers who did not handle dead bodies or human remains.

These data show that despite our best efforts to instil a duty of humanity, we are still far from having reached a situation in which conflicts conform to the moral laws of war. Many factors tend to explain this sad situation, such as the desire to get revenge after a comrade is wounded or killed by the enemy, the dehumanization of the enemy, a lack of leadership and proper training, and a sense of frustration against a very often invisible enemy. There is no doubt that these problems could be partly solved through better training or the establishment of mechanisms that might facilitate the denunciation of criminal activities within the military. However, we can also presuppose that all these measures will never manage to completely prevent massacres from occurring. In fact, we must acknowledge that soldiers' distress after the loss of comrades,[12] or after being exposed to other trauma, can arouse psychopathic violence in ordinary and perfectly sane individuals that the best training and mechanisms will never be able to thwart. Indeed, many contemporary wars have shown this danger, described by Jonathan Shay as 'berserk states of mind' (Shay, 1995), in which soldiers – even the most well trained – enter into a state of reckless frenzy and become unable to distinguish between combatants and non-combatants, and even between comrades and foes. The berserk state of mind allows us to explain what very often appear to be reckless and unjustified murders of civilians committed during warfare, such as the My Lai massacre during the Vietnam War in 1968. On that fateful day, American soldiers went on a killing spree in this village and killed 504 unarmed civilians.[13] This tragic event was the first time these men had a direct encounter with the enemy (at least, this is what they were told), more than three months after their arrival in Vietnam. In the previous weeks, they had been deployed in the mountainous region of Quang Ngai, which was a Vietcong stronghold. During their missions in the thick jungle, these men had only deadly indirect contact with the enemy. On one occasion, they were targeted by non-visible snipers and had to witness one of their radio operators die in agonizing pain after he was shot in the kidney. On another occasion, they stumbled into a minefield and, as it was later recalled by the company commander, Captain Ernest Medina, one man 'was split as if somebody had taken a cleaver right up from his crotch all the way up to his chest cavity' (Lindsay, 2012). In this relatively short period of three months since their deployment, the company had already lost twenty-eight men at the hands of an invisible enemy in gruesome circumstances: a situation that radically changed their mindset. This is what many of the men said about it:

> VARNADO SIMPSON, *rifleman*: Who is the enemy? How can you distinguish between the civilians and the noncivilians? The same people who come and work in the bases at daytime, they just want to shoot and kill you at night-time;

> RONALD GRZESIK, *fire-team leader*: I remember writing a letter home saying that I once had sympathy for these people, but now I didn't care;

> MICHAEL TERRY: A lot of guys didn't feel that they were human beings;

> FRED WIDMER, *radio operator*: Here you are fighting an enemy who doesn't follow the Geneva Convention but you have to abide by it. It's like being [on] a football team where you have to follow the rules to the letter and the other team can do whatever the hell they like. You reach a point where you snap. That is the easiest way to put it; you finally snap. Somebody flicks a switch, and you are a completely different person.' (Lindsay, 2012)

When they were told by Captain Medina on 15 March that they were going to be deployed the very next day in a search-and-destroy-mission to My Lai, the men saw it as an occasion to get revenge for what had happened to their comrades in the previous weeks. As one squad leader said, 'This was a time for us to get even. A time for us to settle the score. A time for revenge' (Lindsay, 2012). Without any mercy, Medina's men performed one of the largest massacre of civilians in contemporary warfare. Some of them later described their psychological state of mind; as Rifleman Simpson said: 'My whole mind just went. It just went ... . I had no feelings, no emotions. Nothing.' Grenadier Thomas Partsch declared: 'As we went through there I don't think the guys realized what they did until after, when it hit them. Maybe some of them did – they were having a ball, but some of the guys were just like in a daze' (Lindsay, 2012).

This event is a clear example of well-trained soldiers entering into berserk states of mind, mainly because of the psychological effects associated with the deaths of their comrades. The same can be said of the murder in Haditha, Iraq, in 2005 of twenty-four unarmed civilians – including a seventy-six-year-old man in a wheelchair and six children – who were shot multiple times almost at point-blank range by US marines, some of whom later urinated on the dead bodies. This tragic event can be explained only through the lens of a berserk state of mind: the shooting was preceded by the death of Lance Corporal Miguel Terrazas after his Humvee was torn apart by an improvised explosive device placed on the road by insurgents. Without excusing the US marines for what they did, it is nonetheless possible to explain this barbarous act by stating that after months of fighting a shadow enemy hiding among the civilians, resulting in the death of one of their comrades, they probably entered a state of frenzy because they wanted to get revenge.

In order to restrict the most repulsive actions during warfare, which are often linked with dysfunctional states of mind in well-trained soldiers, we must not discard technologies that might increase soldiers' moral agency by countering their emotions from a soundly moral standpoint, especially those that can contribute to eliminating feelings directly linked with crimes, namely fear, stress, or frustration. If technologies or medicines can contain soldiers' impulsive aggression and prevent them from entering into an uncontrollable state of mind, then there are solid ethical grounds to use them. As mentioned at the beginning of this book, until now capacity-increasing technologies have mostly been used in order to serve higher military purposes, but we cannot ignore the possibility that they may eventually enforce respect for the laws of war.

Although the prospect of such capacity-increasing technologies is admittedly still hypothetical, it must be noted that the existing devices associated with super soldiers are favourable for better enforcement of the moral rules of warfare. Examples are technologies that help soldiers enhance their vision, their capacity to control their fear and stress, or their alertness on the battlefield. These can play a positive role in respect for the laws of war, as they can ensure a better capacity to distinguish between combatants and non-combatants. Indeed, unpredicted collateral damages are very often the result either of a lack of technological capacity to efficiently understand the environment that surrounds the target or of uncontrolled emotions on the part of combatants, who may end up being confused in the midst of the fog of war.

It must be admitted that the more technologies have evolved over time, the less common such unwanted killings have become.[14] Not long ago, artillery barrage was very often the only option available to an army that wished to seize a town held by the enemy. This type of operation often involved the use of indiscriminate shelling, which ended up costing the lives of civilians. Nowadays, technologies are providing ways to target enemy positions in a more surgical manner, thereby preserving non-combatants from being wounded or killed. Technologies such as multi-focus augmented contact lenses, or the use of transcranial pulse ultrasound mentioned in the previous chapter, would enable the military to boost their alertness and reduce their stress (Tyler, 2010), and would also contribute to a better distinction between combatants and non-combatants by reducing fatigue and incorrect assessment of who is believed to be the enemy – two major elements that can lead to incidents of friendly fire.

Of course, in order for these devices to remain ethical, they should be used in a moral way. If they are misused or diverted in a way that will harm other moral principles of warfare, needless to say we will have valuable reasons to see them as ethically problematic. This issue is a pivotal element at the heart of criticism of the use of unmanned aerial killing vehicles, more commonly referred to as drones, from a moral perspective. This is especially the case for Grégoire Chamayou, for whom these weapons are unethical because they contribute to reversing the logic, already discussed, behind the legitimacy of killing enemy combatants during warfare, and because they often lead to the killing of civilians. While the value of the first point deserves to be recognized, the second is more open to criticism, as Chamayou clumsily confuses the inherent ethical value of the drone with the way it is used.

In his book, Chamayou provides numerous examples of drone attacks that have led to the deaths of individuals who were not affiliated with terrorist organizations. These resulted from a particular type of strike called 'signature attacks' that are authorized on the basis of specific characteristics usually associated with terrorism. In these cases, the decision to kill is made without the precise knowledge of the individuals who are targeted. Since the beginning of the 'War on Terror' following the events of 9/11, these types of strike have represented the majority of drone attacks carried out by the US military (Klaidman, 2012, p. 41). With these rules of engagement, it is clear that mistakes can be very common. For instance, in March 2011, thirty Pakistani civilians who had met for a traditional village assembly were killed by a Hellfire missile launched from a drone. Viewing the event from the sky, the operator thought that they had displayed behaviour peculiar to an al-Qaida training camp. In January 2010, seventeen men who were exercising in the open air were targeted for the same reason (Jeangène Vilmer, 2014). On the basis of these numerous examples of misuse, Chamayou argues that these weapons are fundamentally unethical.

However, such a conclusion is rather problematic as the French philosopher is not making a distinction between the moral *raison d'être* of this weapon and the way it is used. In addition to being a way for the military to ensure its duty of care towards its members by avoiding exposing them to unnecessary risks, a drone also provides a way to fight effectively against individuals and organizations that pose a direct and deadly threat against civilians. While the American rules of engagement are indeed too loose, other states – such as Israel – have been using this method of targeted killing in a more restrictive manner by allowing only 'personnel strikes', that is, operations against individuals whose participation in terrorist activities is beyond any reasonable doubt. The Israeli military has often emphasized the point that an individual will be targeted only if he or she represents a serious threat to others on the basis of reliable information corroborated by several different sources. Moreover, the decision to eliminate the individual will be taken only as a last resort and only if this implies no risk to the civilian population (Guiora, 2004, p. 322). This shows that drones, if used properly with restrictive rules, can be defended from a moral perspective as a means for a state to protect its security and its civilians.

As the reader will realize, this discussion of drones is in line with the way in which capacity-increasing technologies ought to be used on the battlefield. It is not true that if a device or weapon can be defended from strong ethical considerations – as in the case of capacity-increasing technologies – it can be used in every possible way. Other considerations have to be taken into consideration, more especially how it is used and its possible negative consequences on other principles of the rules of warfare. It would be a mistake to condemn these weapons simply on the basis of their misuse. This point will be the focus of the next chapter.

Conclusion

This chapter has attempted to demonstrate that even though capacity-increasing technologies cannot be morally justified in the same way as capacity-restoring technologies, it is nonetheless possible to defend their use from another ethical viewpoint. It has been argued that the military, as an employer, has an obligation to protect the safety, health, and lives of its members in exchange for their obedience. This obligation is common to all institutions and organizations that employ individuals. Of course, the military is somewhat different from other professional organizations, such as governments or construction companies. For the latter, as the work expected from employees is, in itself, not dangerous or life-threatening, these individuals have the right to refuse to fulfil a task if they feel that their working conditions could jeopardize their health or survival, whereas for employees of the military contending with life-threatening situations is integral to the job. Because joining the military is nowadays a voluntary choice in most Western countries, members do not enjoy the same right to refuse to fulfil their duties if they come to believe that it might otherwise lead to risks to their health or their lives. However, this does not mean that the employer consequently has the right to expose them to all dangerous situations and to treat them as expendable goods. The rhetoric used by the military, as well as its inherent operational logic, tends to prove otherwise. This organization also has the moral obligation to do everything in its power to limit, as far as possible, the potential risks to its members' physical and psychological integrity by providing the best possible training before exposing them to dangerous situations and the best available equipment that will allow them to fulfil their professional duties in conditions that engender as little danger as possible, and to refrain from exposing them to unnecessary risks. From this perspective, allowing soldiers to benefit from capacity-increasing technologies that improve their chances of survival or diminish the risks to their health is fundamentally similar to the use of bulletproof vests, helmets, or armoured vehicles for dangerous missions. We might also mention technologies or medicines that allow soldiers to fight sleep deprivation – a typical condition for combatants deployed on the battlefield. Research has indeed demonstrated that this type of fatigue has consequences for people's lives. Indeed, it has been estimated that around 10 per cent of all traffic accidents each year in the United States are connected with drivers' sleep deficit (Kamienski, 2016, p. 265). In the case of soldiers who are armed with deadly weapons and allowed to kill the enemy, it is fairly easy to determine the dangers of sleep deprivation: they may easily confuse a civilian or fellow soldier for an enemy combatant or simply become victims of accidents because of their incapacity to correctly interpret their combat environment. For this reason, technological developments that would increase soldiers' chances of avoiding wounds or death while serving their country are, in themselves, morally necessary (Swearengen and Anderson, 2012, p. 27), and, consequently, it would be a mistake to consider the question of super soldiers as automatically morally dubious simply because they represent an attempt to transform our human nature. Moreover, while these technologies create an asymmetrical dynamic between those who can benefit from them and those who do not have this chance, it would be wrong to regard them as morally problematic so long as they do not create a situation of riskless warfare.

On another level, these technologies open up the prospect of increasing warfare morality by providing soldiers with means to contribute to eliminating feelings that can lead to murderous behaviours against civilians. For the reasons stated previously, the nature of warfare often entails trauma and other psychological effects on combatants, who, despite the best training and their knowledge of the moral rules of war, may end up committing heinous crimes that contribute to exacerbating the terrible reality of combat operations to an unethical degree. Moreover, we cannot ignore the fact that capacity-increasing technologies can contribute to enhancing the morality of warfare – especially the distinction between combatants and non-combatants – by providing tools to soldiers that will ameliorate their judgement and appreciation of threats on the battlefield. In this perspective, it must be acknowledged that these technologies could lead to a moral enhancement of soldiers that could, ultimately, contribute to a greater respect for international humanitarian laws.

Notwithstanding the argument that capacity-increasing technologies are ethically motivated, it is not necessarily the case that we should blindly allow all such technologies. The discussion is, of course, much more complex than this, as the next chapters will show. One of the main hurdles undoubtedly consists of the judicial consequences that might arise from the use of such technologies. The following pages will explore this issue.

Notes

1 These kinds of experiments were preceded by many others. For example, in 1942, the US Army and Navy doctors infected inmates at a Chicago prison with malaria to study

the disease, in the hope of developing a treatment for it. The prisoners were not told that they were going to be infected with malaria. Nazi doctors used this experiment at the Nuremberg Trials to defend what they did in concentration camps. The US Army also tested mustard gas on four thousand soldiers in 1942. These experiments made use of Seventh Day Adventists who chose to become human guinea pigs rather than serve on active duty.

2 In the eyes of many of his contemporaries, Haig was the quintessential example of an inapt military leader. This is the reason why he was nicknamed 'the Butcher of the Somme'.

3 They are some exceptions to this rule. For instance, in 2007 in the US Court of Appeals for the Ninth Circuit, it was determined that soldiers could sue the military if they suffered injuries that could have happened to anyone because of the institution's negligence. This case was raised by Lance Corporal Aaron Schoenfeld, who, while on leave for the weekend, was involved in a car accident on a road in Camp Pendleton when the driver of the car in which he was a passenger crashed into a guardrail that had been badly damaged in a prior accident. The collision severed Schoenfeld's right leg below the knee. He filed a complaint against the government alleging that the military had knowledge of the damaged guardrail and had negligently failed to repair it or warn of its dangerous condition.

4 Jessica Wolfendale and Steve Clarke write in this regard that 'Upholding this duty of care involves a wide array of activities, such as ensuring that military personnel are properly trained, that they are given adequate clothing, weapons, and armor, and that they are physically prepared for the exigencies of combat' (Wolfendale and Clarke, 2008, p. 338).

5 In fact, the United States lost a helicopter after it experienced a hazardous airflow condition. However, none of the men aboard were seriously injured in the crash landing.

6 Even though this strategy had the advantages of simplicity and reduced risk for American soldiers, it was estimated that it would take about fifty thousand pounds of ordnance to destroy the compound and everyone living inside. That power would also have killed civilians living nearby.

7 Such as the Mine-Resistant Ambush Protected vehicles used by the US Army in Iraq and Afghanistan.

8 As mentioned in the Chilcot Report, 'The vehicle was also tested against the RPG 7 [Rocket Propelled Grenade 7] and improvised grenades, as would be expected it does not offer full protection from this type of device.' Chilcot Report, Section 14.1, p. 23, www.iraqinquiry.org.uk/media/246636/the-report-of-the-iraq-inquiry_section-141.pdf (last accessed 27 September 2017).

9 Smith and Others, Ellis, Allbutt, and Others v. Ministry of Defence [2013] UKSC 41, judgment dated 19 June 2013.

10 www.ptsd.va.gov/public/PTSD-overview/basics/how-common-is-ptsd.asp (last accessed 27 September 2017).

11 As previously mentioned, this duty of care is not total for institutions employing individuals in dangerous duties, such as police officers, fire fighters, and soldiers. This logic is integral in the decision in Smith and Others, Ellis, Allbutt, and Others v. Ministry of Defence, where the judges stated accurately, 'The sad fact is that, while members of the armed forces on active service can be given some measure of protection against death and injury, the nature of the job they do means that this can never be complete. They deserve our respect because they are willing to face these risks in the national interest, and the law will always attach importance to the protection of life and physical safety.

But it is of paramount importance that the work that the armed services do in the national interest should not be impeded by having to prepare for or conduct active operations against the enemy under the threat of litigation if things should go wrong. The court must be especially careful, in their case, to have regard to the public interest, to the unpredictable nature of armed conflict and to the inevitable risks that it gives rise to when it is striking the balance as to what is fair, just and reasonable' (par. 100). The duty of care in the military will, of course, have to take into consideration the specificities of what is coined as an omission to protect the soldiers' right to life. For the judges, the balance between the duty of care and the inherent dangers of serving in the military must be established in the following way: 'Close attention must be paid to the time when the alleged failures are said to have taken place, and to the circumstances in which and the persons by whom the decisions that gave rise to them were taken. It will be easier to find that the duty of care has been breached where the failure can be attributed to decisions about training or equipment that were taken before deployment, when there was time to assess the risks to life that had to be planned for, than it will be where they are attributable to what was taking place in theatre. The more constrained he is by decisions that have already been taken for reasons of policy at a high level of command beforehand or by the effects of contact with the enemy, the more difficult it will be to find that the decision-taker in theatre was at fault. Great care needs to be taken not to subject those responsible for decisions at any level that affect what takes place on the battlefield, or in operations of the kind that were being conducted in Iraq after the end of hostilities, to duties that are unrealistic or excessively burdensome' (par. 99).

12 According to the testimonies of veterans of the Vietnam War, this seems to have been the prime factor in the killing of civilians. As mentioned by Philip Caputo, who was court-martialled for the killing of two Vietnamese civilians, 'The war was mostly a matter of enduring weeks of expectant waiting and, at random intervals, of conducting vicious manhunts through jungles and swamps where snipers harassed us constantly and booby traps cut us down one by one. ... At times, the comradeship that was the war's only redeeming quality caused some of the worst crimes – acts of retribution for friends who had been killed. Some men could not withstand the stress of guerrilla-fighting: the hair-trigger alertness constantly demanded of them, the feeling that the enemy was everywhere, the inability to distinguish civilians from combatants created emotional pressures which built to such a point that a trivial provocation could make these men explode [with] the blind destructiveness of a mortar shell' (Caputo, 1977, p. xv).

13 The inquiry led by the US Army established that the number of those killed was 347.

14 Other factors can also play a role, such as a better training and better tactics.

References

Anthony, Sebastian. 2012. 'US Military Developing Multi-Focus Augmented Reality Contact Lenses', *ExtremeTech*, 13 April.

Arkin, Ronald C. 2010. 'The Case for Ethical Autonomy in Unmanned Systems', *Journal of Military Ethics*, Vol. 9, No. 4, pp. 332–341.

Barrie, Allison. 2012. 'Super Soldier Vision Incoming?', FoxNews, 3 May. www.foxnews.com/tech/2012/05/03/super-soldier-vision-incoming.html (last accessed 27 September 2017).

Bell, J. 2008. 'Propranolol, Post-Traumatic Stress Disorder and Narrative Identity', *Journal of Medical Ethics*, Vol. 34, No. 11, e23.

Bernton, Hal. 2011. 'Report: Combat Soldiers Carrying Too Much Weight', *Seattle Times*, 13 February.

Bourke, Joanna. 1999. *An Intimate History of Killing*. New York: Basic Books.

Bowden, Mark. 2012. 'The Hunt for Geronimo', *Vanity Fair*, 12 October.

'Bulletproof Vest Program'. n.d. State of Massachusetts. www.mass.gov/eopss/law-enforce-and-cj/law-enforce/bulletproof-vest-program-overview.html (last accessed 27 September 2017).

Caputo, Philip. 1977. *A Rumor of War*. New York: Owl Books.

Chester, James. 1883. 'Standing Armies a Necessity of Civilization', *The United Service: A Monthly Review of Military and Naval Affairs*, Volume 9, December, pp. 658–666.

Cohen, William and Henry Shelton. 1999. *Joint Statement on Kosovo After-Action Review before the Senate Armed Service Committee*, 14 October.

Feres v. United States, 340 US 135 (1950).

Gardner, Andrew John and John Griffiths. 2014. 'Propranolol, Post-Traumatic Stress Disorder, and Intensive Care: Incorporating New Advances in Psychiatry Into the ICU', *Critical Care*, 18. http://ccforum.com/content/pdf/s13054–014–0698–3.pdf (last accessed 27 September 2017).

Guiora, Amos. 2004. 'Targeted Killing as Active Self-Defense', *Case Western Reserve of International Law*, Vol. 36, No. 2, pp. 319–334.

Hurst, Fabienne. 2013. 'WWII Drug: The German Granddaddy of Crystal Meth', *Spiegel Online*, 30 May. www.spiegel.de/international/germany/crystal-meth-origins-link-back-to-nazi-germany-and-world-war-ii-a-901755.html (last accessed 27 September 2017).

Jeangène Vilmer, Jean-Baptiste. 2014. 'An Ideology of the Drone'. www.booksandideas.net/An-Ideology-of-the-Drone.html (last accessed 27 September 2017).

Kamienski, Lukasz. 2016. *Shooting Up: A Short History of Drugs and War*. Oxford: Oxford University Press.

Ketchum, James S. 2012, *Chemical Warfare Secrets Almost Forgotten: A Personal Story of Medical Testing of Army Volunteers during the Cold War (1955–1975)*. Bloomington, Indiana: Author House.

Khatchadourian, Raffi. 2012. 'Operation Delirium', *New Yorker*, 17 December.

Klaidman, Daniel. 2012. *Kill or Capture: The War on Terror and the Soul of the Obama Presidency*. Boston: Houghton Mifflin Harcourt.

Lavallée, Jean-Luc. 2016. 'Les policiers et pompiers de Québec vont tester un exosquelette', *Journal de Québec*, 9 June.

Lin, Patrick, Maxwell J. Mehlman, and Keith Abney. 2013. *Enhanced Warfighters: Risk, Ethics, and Policy*. The Greenwall Foundation. http://ethics.calpoly.edu/Greenwall_report.pdf (last accessed 8 October 2017).

Lindsay, Drew. 2012. 'Something Dark and Bloody: What Happened in My Lai?' www.historynet.com/something-dark-and-bloody-what-happened-at-my-lai.htm (last accessed 27 September 2017).

Loria, Kevin. 2014. 'Jetpacks Help Soldiers Run at the Speed of Olympic Athletes', *Business Insider*, 12 September.

McNeal, Gregory S. 2011. 'The Aftermath: Why Obama Chose SEAL, Not Drones', *Foreign Policy.com*. http://foreignpolicy.com/2011/05/05/the-bin-laden-aftermath-why-obama-chose-seals-not-drones/ (last accessed 27 September 2017).

Oakley, Rachel. 2015. 'Super Firemen: These Fire Fighting Exoskeleton Suits Give Superhuman Abilities', *Techly*, 22 May.

Office of the Surgeon General. 2006. 'Mental Health Advisory Team (MHAT) IV Operation Iraqi Freedom 05–07 Final Report'. United States Army Medical Command.

Shachtman, Noah. 2007. 'Superchanging Soldiers' Cells', *Wired*, 3 August.

Shay, Jonathan. 1995. *Achilles in Vietnam: Combat Trauma and the Undoing of Character*. New York: Scribner.

Smith and Others, Ellis, Allbutt and Others v. Ministry of Defence [2013] UKSC 41, judgment dated 19 June.

Strawser, Bradley Jay. 2010. 'Moral Predators: The Duty to Employ Uninhabited Aerial Vehicles', *Journal of Military Ethics*, Vol. 9, No. 4, pp. 342–368.

Swearengen, James R. and Arthur O. Anderson. 2012. 'Scientific and Ethical Importance of Animals Models in Biodefense Research'. In James R. Swearengen (ed.), *Biodefense Research Methodology and Animal Models*, 2nd edition. Boca Raton: CRC Press, pp. 27–44.

Tonkens, Ryan. 2015. 'Morally Enhanced Soldiers: Beyond Military Necessity'. In Jai Galliott and Mianna Lotz (eds), *Super Soldiers: The Ethical, Legal and Social Implications*. Farnham, UK: Ashgate, pp. 53–61.

Tyler, William J. 2010. 'Remote Control of Brain Activity Using Ultrasound'. http://science.dodlive.mil/2010/09/01/remote-control-of-brain-activity-using-ultrasound/ (last accessed 27 September 2017)

United States Court of Appeals for the Ninth Circuit. 2007. Schoenfeld v. Quamme and United States of America. No. 05-55126.

Wolfendale, Jessica and Steve Clarke. 2008. 'Paternalism, Consent, and the Use of Experimental Drugs in the Military', *Journal of Medicine and Philosophy*, Vol. 33, No. 4, pp. 337–355.

3

Military capacity-increasing technologies and their consequences

It is wrong and immoral to seek to escape the consequences of one's acts.
Mahatma Gandhi, 1925

The evil of technology was not technology itself, Lindbergh came to see after the war, not in airplanes or the myriad contrivances of modern technical ingenuity, but in the extent to which they can distance us from our better moral nature, or sense of personal accountability.
David McCullough, 1992

As discussed in the preceding chapters, resorting to capacity-increasing technologies for soldiers is not morally problematic. On the contrary, such technologies are tools that can either serve an ethical obligation that the military institution has towards its members or potentially increase the morality of warfare. However, as it is often said, the road to hell is paved with good intentions, and these considerations are not sufficient to assess the rightness of the use of these technologies, as it is also essential to understand their potential flaws from an ethical standpoint. The experiments at Edgewood Arsenal during the Cold War, previously mentioned, offer a good example of this. Some scientists have justified their participation in these experiments as a way to humanize human conflicts. As they saw it, their work was focused on creating chemicals that they considered to be human instruments of warfare. For instance, in 1949 Edgewood's scientific director wrote a classified report entitled 'Psychochemical Warfare: A New Concept of War', in which he argued that 'Throughout recorded history, wars have been characterised by death, human misery, and the destruction of property; each major conflict being more catastrophic than the one preceding it' and that it was possible 'by means of the techniques of psychochemical warfare, to conquer an enemy without the wholesale killing of his people or the mass destruction of his property' (Khatchadourian, 2012). This is also how James S. Ketchum, a psychiatrist who worked at Edgewood between 1960 and 1969, justified his participation in the research programme. In the prologue of his book, Ketchum asks us to imagine a hypothetical situation in which a town is attacked at night with sufentanil, a synthetic chemical so potent that less than half a milligram can quickly produce profoundly incapacitating effects on the central nervous system (Ketchum, 2012, p. i). Immediately after the attack, soldiers would quickly enter the town and medics would hastily check everyone (women and children first, along with the elderly) to make sure they were breathing. Ketchum agrees that such a chemical attack is a breach of international laws and chemical warfare treaties. However, he goes on to argue that it is nonetheless more humane than conventional warfare and hopes that 'eventually, life-sparing drugs, by reducing the acknowledged brutality of conventional warfare, may find acceptance' (Ketchum, 2012, p. ii). For him, the development and use of incapacitating drugs are clearly improvements from an ethical standpoint when compared with conventional weapons that kill people including civilians, who are often victims of collateral damage. In his view, anyone who agrees with the sanctity of human life should consider the legitimacy of using chemical weapons such as those that were developed and tested at Edgewood Arsenal. Although this hypothetical situation appears interesting and not without merit from an ethical standpoint, it nonetheless suffers an important flaw: we can only imagine the tragic consequences if such a chemical were used in a town where civilians were driving their cars, riding bicycles, or swimming in their pools. Their incapacitation would undoubtedly threaten many civilian lives, which would run counter to the norm of discrimination between combatants and non-combatants.

It can be argued that the desire to create super soldiers rests upon similar good moral intentions, such as making them more resistant to effort and stress, providing the means to allow them to make efficient decisions on the battlefield, or increasing their chances of survival. However, this moral intention is not sufficient in itself to justify the use of these technologies, since they also hide implications that might have negative ethical consequences, and these could lead us to reconsider the development and use of capacity-increasing technologies. One of them is clearly associated with the notion of disobedience within the military. Such a pivotal principle has to be cultivated among members of the armed forces, for otherwise war would run the risk of becoming synonymous with pure savagery. It is precisely this fear that capacity-increasing technologies fuel, as they could end up directly or indirectly transforming soldiers into fully obedient killing machines who are unable to exercise any form of moral judgement about what they are ordered to do. Thus the question of super soldiers cannot ignore the possible legal implications of capacity-increasing technologies. One of the main problems is undoubtedly the danger that they might deprive military personnel of their capacity to use their free will and to act autonomously (Wolfendale, 2008). Such an eventuality would have serious judicial consequences. More precisely, the criminal law of some jurisdictions presupposes that individuals cannot be held responsible for crimes they committed if they were in a mental state in which they were unable to distinguish between right and wrong. Bearing in mind that the medicines and technologies associated with the development of super soldiers aim to alter soldiers' psychological faculties, we could ultimately create a situation where war crimes might be unpunished. A second potential legal problem attached to capacity-increasing technologies within the military lies in the possibility that they could alter soldiers' memory and make it impossible for investigators to charge guilty perpetrators, since none of their comrades would be able to testify against them by recalling the events. These two legal problems must be included in a broader discussion about the fundamental moral *jus post bellum* principle of retribution,[1] according to which those responsible for wrongdoings during war should be prosecuted for their crimes. If capacity-increasing technologies open the door to the negation of this principle, it is necessary to fully understand their implications for an efficient transition to peace.

The importance of ethical disobedience within the military

For many people, the military is an institution whose members must blindly obey all tasks they are ordered to perform. This understanding is not in itself inaccurate, and it is true that the military organization depends on instant, unquestioning obedience to those in positions of authority. Without it, the effectiveness of military interventions would be crippled and made impossible, as well as dangerous for the lives of others, since the well-being of fellow soldiers depends on instantaneous compliance with orders. In this sense, granting soldiers the right to question the value and necessity of orders appears complete nonsense.[2] These are the main reasons why insubordination is typically punished in all branches of the military.[3]

In fact, the principle of obedience is a main pillar of the military institution. For von Clausewitz, obedience was seen by the military institution as a way to counterbalance the lack of individual aptitude among soldiers who had a low level of responsibilities (von Clausewitz, 1976, p. 145).[4] Napoléon's *grognards* were probably the last example of soldiers whose complaints were tolerated. However, after the Restoration in 1815, the French army quickly copied the Prussian model, which was famous for its strict training method. This was exemplified by the king's order of 13 May 1818, which stated that, discipline being the main strength of France's armies, the king commanded that all superiors must receive absolute obedience from their subordinates and that all orders must be obeyed without hesitation. Since then, submission to a higher authority has become the norm in all armies of the world, such that the French philosopher Michel Foucault (1975) has compared the military to the penitentiary system.. As a consequence, the training of members of the armed forces during the nineteenth century attempted to stifle any interest in general culture, philosophy, erudition, and critical thinking. On the contrary, the intention was to infantilize soldiers and force them to adopt an unquestioning conformism.

However, the need for soldiers' obedience does not mean that they have to blindly obey all orders they are given. If this were the case, combatants would not have the possibility of refusing to participate in war crimes or in crimes against humanity. This is why soldiers have to obey orders insofar as they are not 'manifestly uenlawful'. In most cases, such a duty is explicitly stated. This is the case for French soldiers, who, according to Decree No. 75-675 of 28 July 1975, 'shall not carry out an order to do something that is manifestly unlawful or contrary to the customs of war, the rules of international law applicable in armed conflicts, or duly ratified or approved international treaties' (Uniform Code of Military Justice, 64 Stat. 109, 10 USC §§ 801–946). In other cases, the duty to disobey unlawful orders is only implicit, as is the case with the American Uniform Code of Military Justice, which states in Articles 90, 91, and 92 that soldiers' obligation consists only in obeying lawful commands. Even though the ban on obeying unlawful orders is not as straightforward as in the case of France, the implications are the same. In either country, a soldier who obeys illegal and immoral commands faces criminal prosecution for his act.

Accordingly, the principle of ethical disobedience can be – and has to be – considered a superior principle, superseding the duty to obey orders. The importance of such a prescription was clear during the Nuremberg Trials, in which Hitler's notorious henchmen were indicted for blindly following the Führer's illegal orders. For instance, Field Marshal Wilhelm Keitel was sentenced for war crimes and crimes against humanity for having signed a series of decrees, such as the Night and Fog Decree, the Commando Order, the Commissar Order, and the orders that the French airmen of the Normandie-Niemen squadron were to be executed upon capture. It was also the fate of other low-ranking officers of the German arm, such as Otto Ohlendorf, a commanding officer of the infamous *Einsatzgruppen*, who was found guilty of murdering more than ninety thousand individuals on the Russian front following orders to do so by his superiors, even though he felt that these orders were 'morally wrong'. In view of his acknowledged unwillingness to exercise moral judgement, the tribunal refused his plea of obedience to superior orders, and he was sentenced to death (Bakker, 1989, p. 73). This principle shows that contrary to what people might think, soldiers are not robots trained *à la Full Metal Jacket*. It is, rather, presumed that they remain capable of using their free will, and they are encouraged to do so.

Of course, this unquestionable obligation to disobey orders that would offend the conscience of every reasonable human being may be theoretically sound, but it is sometimes challenged by reality. One factor that has to be considered is undoubtedly the pressure within the military that may restrain individuals from disobeying such orders or denouncing the illegal actions of their fellow soldiers. One famous example is the case iof Hugh Thompson Jr who, as a helicopter pilot, was a direct witness of the My Lai massacre. During that fateful day, Warrant Officer Thompson saw the men of Company C (Charlie Company), 1st Battalion, 20th Infantry Regiment of Task Force Barker, led by Captain Ernest Medina and Second Lieutenant William Calley, killing women, children, and elderly people in the Vietnamese village. In order to save further innocent lives, he landed his helicopter between the American soldiers and the villagers trying to flee and ordered his two gunners to shoot at their comrades if they attempted to kill the civilians. He then managed to persuade the latter to follow him and ensured their evacuation with two other helicopters. After his return to base, he filled an official report about this war crime to his superiors, who initially managed to cover up the massacre.

While it is undeniable that Warrant Officer Thompson acted in a moral way and should have been praised for his actions,[5] he was nevertheless severely blamed during the congressional investigation. The chairman of the House Armed Services Committee, Mendel Rivers, publicly declared that since he ordered his crewmen to turn their weapons against American soldiers, he should have been the only one punished for the actions in My Lai. He even tried to have him court-martialled. When the public became aware of his actions, Thompson received death threats. Mutilated animals were placed on his doorsteps; he was ostracized by other members of the armed forces and suffered depression in the following years.

This story demonstrates that disobeying illegal orders or denouncing comrades who have followed such orders sometimes has a heavy price for those who make the right decision. The same can be said of Captain Silas S. Soules, who on 29 November 1864 refused to take part in what was later called the Sand Creek massacre, during which a hundred Native Americans (about two-thirds of whom were women and children) were killed and mutilated. After testifying against the officer responsible for these murders, Captain Soules was murdered, presumably in revenge for his denunciation.[6]

Moreover, in view of the manner in which the obligation to disobey illegal orders is formulated, another point to be considered is the extent of permissible disobedience, which may be interpreted in a generous fashion. For instance, if soldiers are not allowed to directly commit illegal actions, can they still be held responsible for facilitating indirectly, but in full knowledge, the perpetration of war crimes or genocide? Two contemporary examples come to mind. The first is the order given to the 370 Dutch peacekeepers stationed at the Muslim enclave of Srebrenica in 1995 to use violent actions only for their own self-defence. As we unfortunately know today, this order – although legal – nonetheless led to the massacre of more than eight thousand Muslim men and boys by the Serb forces of Ratko Mladić.

Refusing to disobey this tragic order had legal consequences for the Netherlands, as the state was held responsible in a 2014 court decision for the deaths of three hundred Bosnian male refugees who were taking refuge in the Dutch mini-compound at Potocari after the fall of Srebrenica (Rechtbank Den Haag, 2014).[7] On 13 July 1995, the Dutch peacekeepers agreed to hand over these men, aged between sixteen and sixty years, to the Bosnian Serb forces for identification, under the terms of the promise by General Mladić that they would be returned (Rechtbank Den Haag, 2014, par. 4.212). According to the tribunal, the Dutch peacekeepers should have known that the men deported from the mini-safe area by Bosnian Serb forces would be killed, because there was strong evidence that the Serbs were committing war crimes.[8] Consequently, the Dutch United Nations battalion acted unlawfully (Rechtbank Den Haag, 2014, par. 4.329).

According to this logic, their knowledge that the men would probably be killed by the Bosnian Serb forces should have led their commander, Thom Karremans, to refuse to obey the Bosnian Serbs' demand and to actively defend these innocent civilians (as he should also have done according to the court's decision). If he had chosen to do so, Commander Karremans and his men would have had to disobey his orders restricting the use of their weapons to self-defence. The Dutch court decision shows that soldiers can be held accountable not only for participating directly in war crimes, but also for facilitating them indirectly.

The example of the Canadian General Roméo Dallaire, the commander of the United Nations Assistance Mission for Rwanda (UNAMIR), the ill-fated United

Nations peacekeeping force in Rwanda, raises the same questions as the case from Srebrenica. In 1994, the world witnessed the massacre of approximately eight hundred thousand Tutsis by Hutu militia in Rwanda and a lack of action on the part of General Dallaire's troops, who did not stop the genocide, a crime that the general knew was going to happen long before to it began. Indeed, as he wrote in his book *Shake Hands with the Devil* (2003), it was obvious many months before the beginning of this sad humanitarian tragedy that a genocide was being organized by the Hutu extremists, and he was informed of this on many occasions. As he recalls in his book, he was first alerted in November 1993 to the discovery of caches of weapons, which led him to realize that 'Something malicious was definitely afoot' (Dallaire, 2003, p. 122). On 20 January 1994, the general was warned by a Hutu informant, a man named Jean-Pierre, that radical members of the Hutu government were planning to eliminate the Tutsis. He sent cables to Kofi Annan, then head of the United Nations Department of Peacekeeping Operations, with this information. His informant had been ordered to register all Tutsis in Kigali, which he suspected to be a method of facilitating future killings. He was also horrified to hear the propaganda of Hutu extremists echoed in the media, notably the *Kangura* newspaper and Radio Télévision Libre des Milles Collines, which explicitly called for the elimination of the Tutsis (Thompson, 2007, pp. 1–12). For this reason, General Dallaire repeatedly asked his superiors for an extension of the United Nations mandate, which forbade him to disarm the militias and to block the Hutus' radio transmissions.[9] However, his requests to conduct deterrent operations against the Hutu extremists were refused by his superiors at the United Nations in New York, and his worst fears became an unfortunate reality in April 1994. By his own account, his inability to persuade his superiors to take pre-emptive actions to prevent the genocide haunts him to this day (Dallaire, 2003, p. 147).

This raises the question of whether General Dallaire should have disobeyed his legal orders by taking actions that, in his eyes, would have prevented the genocide, such as seizing the secret arms caches or dismantling the broadcasting capacities of radio stations that were broadcasting heinous messages. Like the peacekeepers in Srebrenica, he knew that obeying the orders he received would facilitate crimes for which he would have been held criminally responsible if he had committed them himself. There is then a debate over the question of whether soldiers have a duty not only to avoid being direct actors in a crime, but also to avoid being considered as facilitators or accessories to crimes (Caron, 2017).

Another point that complicates the effective use of ethical disobedience in the military is the possible tension between the duty to disobey illegal orders and that to disobey those that are considered to be immoral. Indeed, in some countries, soldiers' voluntary commitment to become servants of the state encompasses the duty to disobey both illegal and immoral orders. This obligation is part of Samuel Huntington's perspective on the limits of soldiers' duty to obey. According to him, when there is a conflict between this duty and basic morality, it is clear that soldiers cannot surrender their moral agency to policy-makers (Huntington, 1957, p. 78), even if the order would not violate any formal norms or rules that the members of the military had previously promised to uphold. In other words, although soldiers are expected to act in conformity with their professional duty, they continue to have full authority to act (in this case, to disobey a legal order) in accordance with moral imperatives and should, therefore refuse to obey certain orders regardless of their legal or illegal nature.

This principle is also enshrined in some military manuals. For instance, the US Army instructs its members to 'do what's right', not only 'legally' but also 'morally'. This logic is also integral to military jurisprudence. The Supreme Court of Canada has argued that soldiers must disobey any order 'that offends the conscience of every reasonable, right-thinking person; it must be an order which is obviously and flagrantly wrong' (R. v. Finta, par. 239). Israeli tribunals have adopted a similar viewpoint, arguing that soldiers must disobey orders that 'wave like a black flag above the order given, as a warning saying: "forbidden"'; an order that is not only detectable by legal experts but also has 'a certain and obvious unlawfulness that stems from the order itself, the criminal character of the order itself or of the acts it demands to be committed, an unlawfulness that pierces the eye and agitates the heart, if the eye be not blind nor the heart closed or corrupt' (Green, 1989, p. 169). However, it must be noted that this obligation extends throughout the military world, unlike the obligation to disobey illegal orders. Of course, because of the subjective nature of the concept of morality, the challenge is to determine when a soldier may rightfully choose to do what is good from a moral perspective even if it implies transgressing a legal norm. One example of this is mercy killing on the battlefield.

Of course, the decision to put an end to another soldier's misery and intolerable pain clashes with other ethical principles of war, more specifically the respect for and non-engagement of anyone who has surrendered or who is out of battle because of sickness or wounds. So the question is whether soldiers who carry out merciful acts which, although still illegal, are based on the logic of ethical disobedience should escape criminal sanctions. However, it can be argued that mercy killings on the battlefield can also find justification in the inherent logic of ethical disobedience (Caron, 2014). Because soldiers are taught and encouraged to use moral judgement in order to act in a way that 'does not offend the conscience of every reasonable, right-thinking person', some soldiers have argued that letting another human being die of intolerable and untreatable pain is immoral and 'pierces the eye and agitates the heart' of every reasonable human being. The case of US Army Captain Rogelio Maynulet is a good example. The officer was ordered to set up a traffic control point near Baghdad in order to capture or kill a high-value target (HVT). A vehicle transporting the HVT passed the checkpoint and a high-speed chase followed. The vehicle transporting the HVT ended up crashing into a nearby house. When Captain Maynulet approached the site, he saw that the driver had a head wound (which was described by the medic as 'the worst he had seen in four years as an army medic'), was making a gurgling sound, and was flapping his arm. Once he had considered the man' state and the medic told him that he was going to die, Captain Maynulet decided to shoot him, since 'he was in a state that [he] didn't think was dignified' and 'it was the honourable thing to do' (United States

v. Maynulet, 2009). In his justification for his actions, Captain Maynulet argued that he acted in a manner consistent with the training he had received prior to his deployment in Iraq.

The same can be said regarding the Goose Green incident during the Falklands War. On 2 June 1982, more than a thousand Argentine prisoners of war were detained near piles of artillery ammunition nearby and, concerned that they might explode, asked and were granted permission from their British guards to move them further. Unfortunately, some of the ammunition exploded, and one man was trapped in the burning blaze, screaming for help. After four to five minutes of unsuccessful attempts to reach him, a medic present at the scene realized that the burning man was fatally wounded and suffering terribly without any chance of being rescued; therefore, he decided to shoot him in order to end his misery. Following a military inquiry, the soldier who committed this act was not sanctioned. Even the Argentine authorities accepted the justification for the killing.

In a real case of mercy killing, soldiers act out of purely ethical concerns. In such a situation, the decision to kill someone is simply motivated by the desire to put a halt to pain that is untreatable and is more than likely to lead to death. This is what has been argued by certain soldiers who have been involved in these incidents (Montgomery, 2005). Their actions were not motivated by malice, but rather by compassion in order to maximize humanity and minimize suffering. In my view, this type of moral reasoning is at the heart of the logic of ethical disobedience. Soldiers are taught, trusted, and encouraged to use reasonable judgement on the battlefield in order to refrain from acting in a way that will offend the conscience of every reasonable person. In cases of genuine mercy killing, this is exactly what soldiers do. They sincerely believe that letting someone die in terrible pain that will almost certainly lead to her or his death is immoral and tends to offend the conscience of humanity more than putting an end to her or his misery by killing her or him. However, when confronted by soldiers who have chosen to give more value to their obligation to disobey immoral orders than to respecting legal orders (in this case, to uphold the rules of war by providing help and protection to wounded combatants), military tribunals have problems in dealing with the tension between these two principles, as the Maynulet case shows.

An order to surrender can also reveal this tension between what soldiers are legally and morally bound to do. The determination of people's obligation to obey or disobey occupies an important position in contemporary debates in political philosophy and has led to significant disagreements, but it is clear that it is founded on what has been labelled the consent theory, that is, a person's obligation to behave in a certain way after committing herself or himself to do so. For Michael Walzer, these individuals are simply bound by the word that they gave upon consenting to the obligation. According to this approach, genuine obligations derive only from people's membership of organizations. However, to create any real obligations, membership implies more than the fact of either being born into the particular organization or silently accepting its rules as a result of socialization. Someone's duty will instead derive from her or his voluntary affiliation to the group

(Walzer, 1970, p. 7). Such a commitment can take numerous forms. As Walzer writes:

> Consenting acts can signify a variety of commitments: our sense of ourselves as members of this or that group, our intention to obey this or that rule or set of rules, our authorization of some persons or group of persons to act on our behalf, our belief in or readiness to stake our lives on 'these truths', whatever they are. We can signify any or all of these things by saying 'yes', or signing our names, or repeating an oath, or joining an organization, or initiating or participating in a social practice. (Walzer, 1970, p. xi)

This is clearly the nature of soldiers' professional obligations: they derive from the covenant which they agree to respect at the moment they join the military, and thus soldiers must submit themselves to a set of implicit and explicit rules. The solemn nature of the oath of enlistment only serves to emphasize the importance of what new soldiers are about to do. Kenneth H. Wenker writes:

> All present stand at attention; the right hand is raised; a relatively high-ranking officer usually administers the oath; the flag is displayed prominently. Where a large number take the oath, there will often be a full parade. Normally those present will wear their dress uniforms. Speeches by dignitaries often help to emphasize the importance of the event. These extraordinary concerns for an action that takes less than a minute to perform serve to impress on all concerned the importance of this promise. On the whole, it would seem that if there ever is an obligation to keep a promise, there would be an obligation to keep this one, due to the special efforts made to solemnize it. (Wenker, 1981)

One of the most important obligations of soldiers in liberal democracies, as a result of their oath, is their inherent subordination to the state, which is the only source of legitimacy. In his seminal book *The Soldier and the State*, Samuel Huntington explains the development of this logic, seeing the nineteenth-century Prussian general Carl von Clausewitz as a pivotal figure in the establishment of this dynamic between the military and the state. More precisely, von Clausewitz was the first to understand the true nature of war as a dual reality: it is simultaneously an autonomous science with its own methods and goals and 'a subordinate science in that its ultimate purposes come from outside itself' (Huntington, 1957, p. 56). For the Prussian general, the ultimate purpose was an act of force to compel one's adversary to do one's will, and thus war is nothing but the continuation of policy by other means. According to this definition, if the activity of war is simply a means for states to pursue their external political goals, members of the armed forces are also tools for those who determine the nature and extent of the violence that can be used to satisfy specific political ends. In this dynamic, the former must submit to and obey the will of the latter.[10] This perspective is now the norm within liberal democracies and, as stated by the French legal theorist Raoul Girardet, 'The military must be a passive instrument in the hands of the government, which excludes the possibility for soldiers to refuse to obey the orders given to them by their statesmen' (Girardet, 1960, p. 5, my translation). Since the nineteenth century, this principle has been a core element of warfare and morality classes at the Saint-Cyr military academy, where future officers in the French army are told that 'the loyalty of the army and its dedication to the legal government must be absolute' (Girardet, 1999, p. 548, my translation).

For Huntington, this principle covers a wide variety of cases. For instance, he notes that even in situations when the soldier does not agree with the decision to go to war, he or she must nonetheless accept the policy-makers' decisions. It must be noted that this dynamic between soldiers and state is now integral to our interpretation of Just War Theory: whereas soldiers can be held responsible for their own personal misconduct during a war (for instance, if they kill civilians or mistreat prisoners of war), most cannot share the moral blame for engaging in an unjust war of aggression and crimes against peace because that decision is made solely by policy-makers.[11]

As it has been argued, although soldiers' obligation to obey may be strong and extensive, as highlighted by Huntington, it is not absolute: they also must formally pledge in various ways to refrain from obeying what are generally labelled 'manifestly unlawful orders' and, in some military contexts, those that are thought to be immoral. Charles de Gaulle's decision to resist his country's surrender to the Nazis in June 1940 shows how the tension between the duties to obey legal orders and to disobey immoral orders can create a crisis among soldiers.

When discussing soldiers' obligation to lay down their arms, the first question that must be addressed is whether the policy-makers who give the order have the legitimacy to do so. Indeed, it is clear from Huntington's perspective that soldiers are bound to obey policy-makers, who are indeed the lawful representatives of the state's authority. In this sense, individuals who usurp power by acting outside the rule of law do not have the moral legitimacy to expect members of the military to obey them (Huntington, 1957, p. 77).

We must therefore consider whether this was the case for General de Gaulle in June 1940 if we wish to evaluate whether his desire to continue fighting was a legitimate case of disobedience stemming from his voluntary pledge to be subordinated only to legitimate representatives of the French government. Were the representatives of the French government of the time the lawful embodiment of the state's sovereignty, or were they usurpers whom de Gaulle and his comrades were not bound to obey?

It must be acknowledged that the French government did indeed comprise the lawful representatives of the state and that the decision to surrender was in keeping with the constitutional order of the time (Girardet, 1999, p. 548). Under the constitution of the Third Republic, the President of the Republic, who was elected by the members of the National Assembly and the Senate, was responsible for naming the prime minister (called the *Président du Conseil*), who was in charge of administering the country. When France was attacked by the Germans, the government was led by Paul Reynaud, who had been officially nominated by President Albert Lebrun on 22 March 1940. However, when Reynaud realized that defeat was imminent and that the other members of his government were unwilling to continue fighting the Germans from the French colonies of North Africa, he decided to resign on 16 June. President Lebrun then replaced him with Marshal Philippe

Pétain, who publicly announced the next day that he had asked the Germans for a ceasefire. France officially surrendered in a humiliating ceremony on 22 June.[12]

This means that at the time when General de Gaulle decided to continue fighting the Germans alongside the British on 18 June, he purposely refused to obey a government whose actions were legitimate according to the constitutional principles of the Third Republic and whose decision to surrender could not be criticized from a legal point of view. From this perspective, de Gaulle's actions were not in accordance with his oath as a member of the French military. Accordingly, his decision to disobey his country's order to surrender cannot be considered to be a legitimate resolution that stemmed from his explicit commitment to serve the state. Indeed, de Gaulle's oath as a soldier should have ensured his obedience to Marshal Pétain's decision to accept France's defeat. Legally speaking, his subsequent convictions, the first by a military tribunal on 4 July 1940 for his refusal to obey and for inciting other members of the French military to do the same,[13] and the second a month later for treason, attempting to jeopardize the security of the state, and desertion in a time of war,[14] appear to be perfectly justified within the realm of Walzer's consent theory. His status as a rogue soldier was in fact well known among the military community. Even British officers appointed by the War Office warned other French soldiers who wanted to join de Gaulle that they would be regarded as rebels in the eyes of the military community.[15] De Gaulle himself acknowledged that he was challenging the traditional logic of military obedience. He accused his fellow soldiers who believed that they were acting rightly by remaining faithful to the legal order of the Vichy regime of being prisoners of a 'false discipline' ('fausse discipline') (de Gaulle, 1954, p. 92). Thus his controversial decision to continue to fight the Nazis was contrary to the true duty of a French soldier.

Moreover, in itself, the armistice treaty with the Germans did not lead French soldiers to commit what at the time were illegal actions. It is true that Marshal Pétain's peace led to the deportation and death of thousands of French Jews in the Auschwitz gas chambers, and in this sense, French soldiers could be considered, like the peacekeepers in Rwanda and in Srebrenica, to have been facilitators of a crime against humanity since they agreed to lay down their weapons. But the interdiction of these crimes was not *per se* illegal according to the International Law of the time since, the use of slave labour, a violation of what was part of the international norm was impossible to predict in June 1940. Indeed, the establishment of the slave labour policy was an unplanned war contingency to compensate for the lack of a German workforce caused by sending an increasing number of German soldiers to the Eastern front.[16]

However, no one would now dare to criticize de Gaulle's decision to disobey Marshal Pétain's government by inspiring his countrymen to fight a regime that was the incarnation of pure evil. This obvious conclusion emerges from the obligation that some soldiers have to disobey immoral orders. The treatment of the Jews in Germany from 1933 until 1940 should have led any reasonably informed individual to see the true immoral nature of the Nazi regime. General de Gaulle's disobedience to Marshal Pétain's order to surrender is more justifiable in the light of soldiers' obligation to disobey immoral orders. Indeed, one could argue that although the order to surrender was legal, it was nonetheless immoral because it meant that

France had to submit to the rule of a notorious inhuman regime whose actions were in absolute contradiction to Kant's categorical imperative. It is the immoral nature of the Nazi regime that allows us to justify de Gaulle's disobedience, even though his decision was at the time technically illegal according to a purely legalist view of soldiers' obligation to obey their policy-makers.

The de Gaulle example is far from an exception in the history of warfare. The surrender of the Confederate army of General Robert E. Lee in the American Civil War in 1865 provides a good example of what should prevail when legal and immoral orders conflict. For Lee's soldiers, who for four years had fought bravely in horrific battles against their brothers in the Union army, their surrender was no doubt a terrible blow. Nevertheless, it would be rather weak to argue that their humiliation as vanquished warriors was an immoral act that would offend the conscience of every reasonable human being, especially given that the Confederate army's defeat brought an end to slavery in the United States. The defeat actually served a higher moral end, and the Confederate army's eventual resistance to it would have been condemnable from not only a legal perspective but also from a moral one. However, let us consider the hypothetical situation in which the Confederacy had won the Civil War against the Union. After a series of defeats and the fall of Washington at the hands of General Lee, President Lincoln would have been the one asking for peace at McLean House in Appomattox on 9 April 1865. There is no doubt that Union soldiers would have had to lay down their weapons, and that this order would have been perfectly legal according to the principle of the subordination of the military to the state. Nevertheless, it would have been highly immoral, because it would have led, as would have been well-known at the time, to the establishment of an inhuman regime of cruelty and humiliation through the spread of slavery throughout the United States, violating our Kantian moral duties and resulting in a situation that would have pierced the eye, agitated the heart, and offended the conscience of every reasonable, right-thinking person. Despite being legal, the Union army's capitulation would have been more troublesome from a moral perspective and, according to what has been argued in this short section, the Union soldiers should have resisted it.

Although the debate about ethical disobedience is interesting and deserves a more thorough investigation, it is not my intention to discuss it further in this book as it would be an unnecessary digression from the topic of super soldiers. The foregoing discussion, however, has the advantage of showing the complexity of this notion within the military and how thoroughly we need to study it. The remainder of this chapter will highlight one significant consequence that these capacity-increasing techniques may have on soldiers' duty to disobey illegal actions, the risk of losing their capacity to determine freely whether or not an order is immoral.

The importance of preserving soldiers' moral autonomy: the judicial consequences of capacity-increasing technologies and their implications on *jus post bellum*

The previous discussion has shown that, in order to avoid the perpetration of war crimes, it is essential for soldiers to be able to disobey illegal and immoral orders that they may be asked to follow. Of course, in order for them to do so, soldiers' moral agency must be preserved at all cost. For this reason, the preservation of soldiers' moral autonomy should therefore be a non-negotiable principle, and technologies that have the objective of entirely or partially depriving combatants of this pivotal ability in order to increase their obedience should not be tolerated. In fact, the alteration of soldiers' moral abilities would return us to a pre-Nuremberg situation with regard to the respect of international humanitarian laws (Leveringhaus, 2015, p. 148). The notion of technologies that completely or partially control soldiers' minds and make them fully obedient – even to the most heinous and immoral orders – is simply terrifying, and, moreover, such technologies are illegal under international law. As the philosopher Patrick Lin argues (2013), soldiers are weapons or instruments of war, and there are good reasons to include them under Article 36 of the Geneva Convention (Additional Protocol I of 1977), which specifies that 'In the study, development, acquisition or adoption of a new weapon, means or method of warfare, a High Contracting Party is under an obligation to determine whether its employment would, in some or all circumstances, be prohibited by this Protocol or by any other rule of international law applicable to the High Contracting Party.'

According to this international norm, which has been implemented in order to maintain a certain sense of humanity and restraint in situations where killing other human beings is tolerated, states have the responsibility to assess whether their weapons and means or methods of warfare conform to the principles of discrimination,[17] proportionality,[18] and the prohibition of superfluous injury or unnecessary suffering.[19] With regard to super soldiers, the first principle is clearly the one that is the most relevant to the duty to refrain from committing war crimes against combatants and non-combatants. The legitimate fear in this area is that the use of capacity-increasing technologies might be hampered negatively through the use of certain drugs, medicines, or equipment that would transform soldiers into inhumane killing machines who would not hesitate to kill non-combatants.[20] If technologies with this terrible potential were ever used on the battlefield, they would clearly be in breach of this article of warfare. Accordingly, the manner in which these technologies may directly or indirectly affect soldiers' moral agency cannot be ignored, especially in view of their damaging effects on another set of normative moral principles associated with justice after war ends. This may be a concern regarding the development and use of the transcranial pulse ultrasound technology mentioned earlier. If its aim is simply to relieve soldiers from battlefield stress and boost their alertness, then it can legitimately be considered ethica. We would, however, have to change our perspective if it led the military to take control of soldiers' minds by transforming them into entirely obedient agents without any kind of moral agency, as with *Star Trek*'s 'collar of obedience'. If this was the case, soldiers would lose their capacity to critically reflect on their orders and to possibly disobey them if they believed them to be immoral or illegal.

However, this discussion should not simply be limited to capacity-increasing technologies that have direct and intentional negative consequences on soldiers' moral agency. It should also consider their unintentional impact on soldiers' behaviour and moral agency, as well as the legal consequences this might entail. We can easily imagine a type of capacity-increasing technology whose primary aim would be in line with the military's duty of care, for example by reducing soldiers' vulnerability and exposure to life-threatening situations or by limiting their stress and anxiety on the battlefield, but which would unintentionally harm their moral agency nonetheless. If, as a result of such technologies, soldiers end up committing war crimes against civilians, how should we deal with them? As Alex Leveringhaus has rightly suggested, responsibility can be assigned only if the soldier made an informed decision to relinquish a part of his autonomy when he or she decided to use the technology. In criminal law, such a situation would be very similar to voluntary intoxication, which means intoxicating oneself in the full knowledge that one will become impaired. In many judicial systems around the world, voluntary intoxication is not ordinarily considered as a defence for a crime. For instance, the judges of the Supreme Court of Canada stated in 2011 that 'A malfunctioning of the mind that results exclusively from self-induced intoxication cannot be considered a disease of the mind in the legal sense ...' (Supreme Court of Canada, 2011).[21] This prevents individuals who use drugs or alcohol to plead a reduced sentence for their felonies. Even though their judgement was affected when their crimes were committed, the fact remains that they consciously took inebriating substances knowing the impact they might have on their judgement. Even in cases where these substances may have produced a psychotic reaction, individuals who are voluntary intoxicated are judicially unable to plead mental illness as a defence.[22] Thus the soldier who ended up committing crimes after he decided to use a capacity-increasing technology could be held responsible for his mistake only if he or she consented to its use in full knowledge of its possible side effects on her or his judgement. Yet what if the soldier is not told about these possible side effects? How should we therefore deal with their crimes?

This possibility raises the spectre of crime committed during involuntary intoxication, which refers to a criminal act performed while under the influence of intoxicating substances ingested involuntarily that render the individual incapable of understanding the nature of the acts committed. According to the general jurisprudence, an individual in this state should not be held responsible for her or his actions, since the involuntarily intoxicated person is normally regarded more as a 'victim' than as an offender (Quebec Court, 2011; Court of Appeals of California, 1983; City of Minneapolis, 1976; District Court of Appeal of Florida, 1998).[23] If soldiers are indeed forced to take capacity-increasing technologies that may lead them to commit crimes on the battlefield (without being aware that this might be one of the results), then it would certainly be possible to avoid the legal consequences for their actions, as their crime would result from involuntary intoxication. After all, as mentioned by Stephen E. White, in order to prove the guilt of a war criminal under international law, a prosecutor must prove both the *actus reus* and the *mens rea* of the individual. More precisely, this individual will be sentenced insofar as the crime 'resulted from a voluntary act or wilfull omission' and if he or she 'possessed a culpable state of mind at the time of the killing' (White,

2008, pp. 194–195). In cases of involuntarily intoxicated individuals these criteria are lacking, and this might negatively affect the fundamental principles of *jus post bellum*, as these individuals ought to escape prosecution; this is a possibility that could harm the establishment of a just peace in the aftermath of a war, as the following pages will discuss.

If the risk of altering soldiers' minds without their knowledge or their consent through capacity-increasing technologies should cannot be ruled out, the same danger also applies for soldiers who take a variety of medicines whose interaction with one another could lead them to perform illegal actions. For example, let us assume that soldiers have to take various drugs which, individually, do not impair their moral judgement and their psychological capacities. However, in combination they may cause an interaction that could lead to psychotic behaviours and unlawful decisions. Such a situation would probably also fall in the domain of involuntary intoxication, and, according to the judicial norm, soldiers who commit crimes or illegal actions in such circumstances could not be held criminally accountable. This fantastic scenario may actually not be entirely hypothetical, as the case of US Army Staff Sergeant Robert Bales, who shot and stabbed sixteen civilians in a killing frenzy, most of them women and children, near Kandahar in Afghanistan 2012, raises some legitimate doubts. Sergeant Bales is believed to have taken an anti-malarial medicine called mefloquine during his deployment in Iraq and Afghanistan. This drug has been commonly used for the last thirty years by the US Army to prevent and treat malaria among soldiers and workers deployed abroad. However, it is not without its side effects. According to documents from the manufacturer, Roche, mefloquine 'may cause long-lasting serious mental-health problems' that can persist for months after the drug has been taken (Bernton, 2013). A former army epidemiologist has stated that these side effects include hallucinations, suicidal thoughts, and psychotic behaviour (Shapiro, 2013). The effects seem to be well known among US soldiers, and it has been reported that some of them have developed a liking for it 'because it gave them vivid dreams' (Bernton, 2013). Moreover, the manufacturer has warned against prescribing it to anyone who has previously suffered a seizure or brain injury and it has also been reported as associated with violence towards others (Moore, Glenmullen, and Furberg, 2010). It must be noted that during his previous deployment in Iraq, Sergeant Bales had suffered head injuries. We will never know whether mefloquine actually played a role in his killing spree, as his medical records in Afghanistan were incomplete and he was unable to remember the names of the medicines he took.[24] One can only wonder whether his crimes were triggered by intoxication caused by the interaction of a drug with a pre-existing medical condition: an intoxication that would have to be considered to be involuntary.

These cases show that capacity-increasing technologies create the potential for lowering moral accountability on the part of combatants. As mentioned previously, despite the fact that they can be seen as a moral necessity from the perspective of an institutional duty of care, their use raises questions that cannot simply be swept under the rug. The main problem may be that beneficiaries of capacity-increasing technologies could be prosecuted for actions committed while involuntarily intoxicated – which would be rather unjust – or could avoid prosecution because prosecutors would conclude that their *actus reus* and *mens rea* were lacking at the time of the offence. If the latter became the norm, this might very well 'encourage the development of an entire classes of weapons that the state could use to evade criminal penalties for even the most serious types of war crimes' (White, 2008, p. 206).

Yet these implications would remain incomplete if they were not linked to another component of Just War Theory: *jus post bellum*. While Just War Theory focuses on the moral criteria that allow a state to start a war (*jus ad bellum*) and the way wars should be fought (*jus in bello*), it would be a mistake to ignore or neglect how a just peace has to be established in the aftermath of a conflict. This is what *jus post bellum* is about. As Larry May has pointed out, the principles of *jus post bellum* 'are normative in that they are moral norms' (May, 2012, p. 5). They thus have to be considered to be just as important as the institutional duty of care and soldiers' ethical disobedience. For May, this third stage of Just War Theory consists of six essential normative components: retribution, reconciliation, rebuilding, restitution, reparation, and proportionality. Only the first two of these need to be considered in relation to capacity-increasing technologies.

Retribution implies that those responsible for breaching the rules of war should be prosecuted for their wrongdoings as a matter of justice, but also in order to end the victimization process by rehabilitating the victims and reconciling the opposing nations. From a theoretical perspective, this principle is not solely restricted to crimes committed during wartime (such as the killing of civilians), but also applies to those who have initiated an unjust war. More precisely, a state leader who launches an unjustified war of aggression either does not use proportionate means of action, or does not use war as a last resort. This category of individuals is not restricted to tyrants and dictators, such as Hitler, but may also include democratic leaders. Since the publication of the Chilcot Report in July 2016, many individuals have concluded that this notion should also include democratic leaders such as Tony Blair and George W. Bush, who started the war in Iraq on the basis of wrong assertions. We must admit that this conclusion has some validity. Retribution is generally thought of as having a deterrent effect on individuals who would otherwise not hesitate to perform unjust actions during wartime or initiate unjust wars.

For its part, the principle of reconciliation 'involves a process of returning previously warring parties to a point not only where they do not engage in violence toward each other but also where there is sufficient trust so that a robust and just peace can be attained ...' (May, 2012, p. 86). While retribution for individuals who have breached the rules of war can sometimes have negative consequences on reconciliation and the establishment of long-lasting peace, it must be conceded that such a clash between these two normative imperatives should always be an exception. Sacrificing justice for the sake of peace should be allowed only if the trial of war criminals would result in further violations of human rights. Such an exception should be considered only for a handful of individuals whose popularity among their people might trigger a sense of revenge and a desire to resume combat operations. A good example of this would be the former President of Serbia Slobodan Milosevic, who was suspected of war crimes, including genocide and crimes against humanity, during the wars in Bosnia and Croatia (although he died before the end of his trial in 2006); nonetheless, he was integral to the Dayton Agreement of 1995. Despite his assumed criminal and moral responsibilities for what happened during the war, his presence at the negotiation table was seen as the only way to ensure long-lasting peace in the Balkans. There were fears that his removal from office might destabilize domestic and regional peace and that his supporters might use his indictment as a reason to resume the war. This view that retribution can be sacrificed for the sake of peace and reconciliation has been shared by many individuals, who have argued that post-war reconciliation is more important than the punishment of those who have committed wrongdoings during wartime. For example, the Swiss philosopher Emer de Vattel wrote the following in the eighteenth century:

> Strict justice should not always be insisted on: peace is so advantageous to Nations that they are so strictly under an obligation to cultivate it, and to procure the return of it when it has been lost by war, that when obstacles, such as those above mentioned, are met with in the execution of a treaty of peace, the parties should lend themselves in good faith to all reasonable expedients, and should accept an equivalent, or a compensation, for the act which cannot be performed, rather than annul a peace treaty and renew the war. (Vattel, 1916, pp. 360–361)

This desire to make reconciliation prevail over retribution does not apply to soldiers whose capacities have been increased by technologies. In their case, retribution would not negatively affect the establishment of reconciliation between the former warring parties. On the contrary, their trial would be beneficial for this objective, as it would show the defeated party that even the victors punish wrongdoers for crimes committed during wartime. According to May, the obligation to put these individuals on trial is unquestionable.[25] The subtlety about the question has more to do with whether they should be tried during or after conflict (May, 2012, pp. 75–80).

In view of this normative position, we may wonder how it is possible to have retribution in cases when soldiers have committed war crimes on the battlefield while under the influence of medicines or technologies that directly or indirectly altered their moral agency. In other words, if soldiers commit wrongdoings because of involuntary intoxication or as an unintentional result of the drugs and technologies they have agreed to use, this may create legal loopholes that may make it impossible to try them. Retribution as a condition of reconciliation may be harmed in this context, and would contribute to leaving open the scars of war between former enemies, since individuals who had taken part in crimes during the conflict would remain free from prosecution.

Moreover, the ability to charge a soldier for alleged war crimes often depends on the capacity to find witnesses who are willing and able to testify. Thanks to Hugh Thompson Jr, this is how the My Lai massacre was denounced. However, capacity-increasing technologies might render this possibility a mere illusion. Such is the danger with technologies and medicines that tend to alter soldiers' memory. Once again, this desire relies on good intentions and is in accordance with the army's duty of care towards its members. It is useful to remember that this includes its obligation not only to deploy all necessary means in order to prevent its members from being harmed or killed on the battlefield, but also to make sure that they will be able to resume normal life after their departure from the army. Thus fighting against the prevalence of PTSD is a moral obligation on the part of the military. While only 8 per cent of the general population is affected by this condition, the proportion increases to 30 per cent for individuals who currently are or have been enlisted in the military (US Department of Veterans Affairs, n.d.) – a problem that cannot be ignored. Approximately 75 per cent of former soldiers suffering from PTSD also have substance abuse disorder, and PTSD is also associated with a significantly higher risk of suicide than in the general population –45 per cent higher according to the Ombudsman of Canadian Ministry of National Defence (National Defence and the Canadian Armed Forces, 2016) – as well as violent outbursts. Needless to say, as argued by the US President's Council on Bioethics (*Beyond Therapy*, 2003), PTSD makes daily life hard or even impossible.

For these reasons the army has shown a legitimate interest in the development of treatment that can prevent this affliction. Nowadays, corticosteroids and propranolol are seen as useful in preventing the problem, as they can influence the encoding of memory. More precisely, when taken immediately after a traumatic event, they are able to blunt the victim's memory. Few would argue that these drugs do not satisfy the military's duty of care for its members.[26] However, there is a need to consider the consequences of such medicines, especially the legal ones, and their implications for retribution. Let us assume a hypothetical situation: while on a mission in a remote village, a member of a small unit of soldiers decides to rape or kill unarmed civilians. His colleagues, fearing that they might suffer traumatic memories of what they have witnessed, decide to take propranolol and, as a consequence, none of them can later remember precisely what happened in the village. In such a situation, seeking retribution might be extremely difficult, if not impossible, since no one would be able to testify about what happened. An inquiry would therefore not be able to establish the precise responsibility of the soldiers in the massacre. In a way, the use of the medicine would lead to an unintentional tampering with the evidence of a crime, which is a felony in many jurisdictions. For legal reasons, therefore, the desire to interfere with the memories of trauma survivors and witnesses is questionable as it can have tremendous negative consequences on the quest for justice.

This discussion would be incomplete without addressing the possible impacts these drugs might have on people's obligation to remember. As mentioned in the report of the US President's Council on Bioethics, people who have access to these medicines may become emotionally desensitized to traumatic events such as crimes. For example, psychologically sane combatants who witness unlawful acts on the battlefield they are, with good reason, morally disgusted by them and see them as unacceptable crimes. However, if the same combatants have access to propranolol and similar drugs they may develop an insensitivity to such heinous acts. This is the possibility raised by members of the Council on Bioethics:

> In manipulating [the soldier's memory], he risks coming to think about that murder as more tolerable than it really is, as an event that should not sting [those] who witness it. ... If psychologically, the murder is transformed into an event our witness can recall without pain – or without *any* particular emotion – perhaps its moral significance will also fade from consciousness. If so, he would in a sense have ceased to be a genuine witness of the murder. When asked about it, he might say, 'Yes, I was there. But it wasn't so terrible'. ... Armed with new powers to ease the suffering of bad memories, we might come to see all psychic pain as unnecessary and in the process come to pursue a happiness that is less than human: an unmindful happiness, unchanged by time and events, unmoved by life's vicissitudes. (*Beyond Therapy*, 2003, pp. 258–259)

Therefore the capacity to avoid the power of our own judgement for our actions would be the direct consequence of such medicines. Since individuals would be in a position to avoid having to deal with the harsh judgement of their consciences, not only would they tend to feel less responsible for their actions, but they might well believe that crimes are easier to deal with if they are able to avoid suffering the psychological repercussions of their misdemeanours. Without this deterrent for unlawful and immoral actions, we might very well damage a main component of ethical behaviour. Moreover, we cannot ignore the fact that this possibility is also problematic for the community as a whole. For example, the suffering of the Jews during the Second World War or of the Bosnians in Srebrenica, although traumatic for the victims, was nonetheless a global necessity, as these instances are useful reminders of what humanity should avoid repeating in the future. Traumas contribute to people's sense of empathy. For instance, experiencing the death of a close relative will encourage compassion and sympathy in the victim for other people who are facing such a loss. The same logic applies for individuals who lose their jobs, for someone who is humiliated in public, for a woman or a child suffering from domestic violence, or for someone whose spouse is having an extra-marital affair. There is indeed a strong likelihood that a world in which such traumas lost their significance would also negatively affect our sense of empathy towards others. The use of these capacity-increasing technologies might transform normal human beings into cold and emotionally neutral machines. Such a risk will be discussed more at length in the last chapter of this book.

Despite the good intentions that can justify the use of capacity-increasing technologies, it is clear that they clash with other significant moral imperatives that might impair soldiers' ethical behaviour, as well as the imperatives of *jus post bellum*. The next section therefore suggests general guidelines for distinguishing between capacity-increasing technologies that should be tolerated and those we should avoid using.

General considerations concerning the use of capacity-increasing technologies

The debate about the acceptable and unacceptable ways of increasing soldiers' capacity centres on the opposition between different moral imperatives. On the one hand, and as discussed in the previous chapter, the military has a duty of care towards its members and must, accordingly, deploy all means to protect their life and health. Moreover, we cannot neglect the fact that capacity-increasing technologies can also help states to increase the morality of warfare. On the other hand, there is also a need for soldiers to remain autonomous moral agents able to prioritize superior ethical concerns over their duty to obey as well as the normative implications of justice after war ends. The challenge is to establish which moral imperatives should dominate. Although the duty of care is a moral necessity, the implications of capacity-increasing technologies are far too important to be ignored, as it is hardly conceivable that soldiers should be deprived of their capacity to disobey illegal and immoral orders or be able to escape retribution for their crimes. Their moral responsibility appears to be a higher moral imperative. However, this moral dilemma is not as important as it appears at first sight, as there are other means for the military to ensure its duty of care. Capacity-increasing technologies are only one option that the military has at its disposal to protect the life and health of its members or to enable its service personnel to resume normal life upon their return from the battlefield.

For this reason, extreme caution should be taken with regard to technologies that affect soldiers' minds. For example, we might mention the development of more advanced helmets in order to stimulate soldiers' brains (Homeland Security News Wire, 2010). This technology, developed by scientists at Arizona State University, is by no means new and is currently used to treat patients suffering from an advanced form of Parkinson's disease and from depression, through the transmission of 'transcranial pulsed ultrasound'. In view of its potential military applications, it is not surprising that DARPA has shown an interest and has granted funding for scientists to develop applications for US soldiers. DARPA's goal would be to use it as a way to relieve stress, anxiety, fatigue, and pain during combat missions, as well as to boost awareness.[27] Of course, if this technology was ever to be used in the military, it would undoubtedly contribute to diminishing soldiers' risk to their lives, as lack of concentration, fatigue, and stress often result in mistakes that cost lives.[28] In this respect, it would respect the military's duty of care.

However, the fact that the military might be able to remotely control and influence soldiers' minds on the battlefield is problematic and could have both intended and unintended consequences on the perpetration of illegal crimes. Moreover, making soldiers more obedient and less anxious during war could also lead to a violation of the military's duty of care by encouraging them to fulfil unacceptably dangerous missions without provoking any hesitation or questioning on their part. Because of this fear, this technology could bring us back to situations where the military can treat its members as cannon fodder without any regard for their well-being; this is a problem with which it has been historically linked, as the Zulus' use of traditional plants with psychotropic effects during the nineteenth century proved (Kamienski, 2016, p. 86). Finally, from a tactical perspective, this technology may prove to be as sound as it appears, because enemies could potentially take control of the devices at a distance and incite the soldiers to turn their weapons against their comrades or, even worse, against innocent civilians. Such a possibility would be relatively easy to realize, and not very costly. As reported in 2009, Iraqi insurgents were able to intercept the live feed of US Predator drones through the use of a mass-market software program available for as little as $25.95 on the Internet (Mount and Quijano, 2009).

In view of these possibly negative implications, we ought to wonder if it is possible for the military to develop alternative ways to increase soldiers' awareness and reduce their stress on the battlefield. It seems to be the case that these technologies are not indispensable and could be replaced with other, traditional methods that do not entail the potential problems discussed in this chapter, such as more realistic fighting training or a better rotation system among the troops deployed on combat operations. Of course, these traditional methods may be less effective for treating these problems than capacity-increasing technologies, but in light of the potential problems that may arise from their use, a principle of precaution should prevail and that less dangerous traditional methods ought to be used.

The same logic applies to memory-blurring drugs, such as those used for the treatment of PTSD. Again, although it is difficult to deny their merits for those affected by this terrible syndrome, the military has at its disposal various ways of helping combatants suffering from it. For instance, the US Department of Defense has for some years been using an experimental treatment called 'virtual Iraq' in which veterans work through their combat trauma in a computer-simulated environment that has many similarities with the popular video game *Full Spectrum Warrior* (Halpern, 2008).[29] This experiment uses a process called 'prolonged-exposure therapy' and rests on the following logic:

> It is a kind of cognitive-behavioral therapy, derived from Pavlov's classic work with dogs. Prolonged-exposure therapy … is at once intuitively obvious and counterintuitive: it requires the patient to revisit and retell the story of the trauma over and over again and, through a psychological process called 'habituation', rid it of its overwhelming power. The idea is to disconnect the memory from the reactions to the memory, so that although the memory of the traumatic event remains, the everyday things that can trigger fear and panic, such as trash blowing across the interstate or a car backfiring – what psychologists refer to as cues – are restored to insignificance. The trauma thus becomes a discrete event, not a constant, self-replicating, encompassing condition. (Halpern, 2008)

The military also has at its disposal more conventional treatments, such as the use of individualized trauma-focused cognitive behavioural therapy and other forms of psychological help; these can also help to end the demonization of this problem and the fear that confessing one's psychological distress will have implications for his or her career – an unacceptable weakness for professional soldiers. Of course, this strategy could be more costly for the military as it would involve the recruitment of hundreds, if not thousands, of psychiatrists and psychologists. However, when it comes to the duty of care and treating individuals in a Kantian way, financial cost should never be an acceptable objection. From that deontological perspective, this is not a moral dilemma, but rather a political choice that, of course, does not have the same moral significance.

Thus the price of allowing the military to use devices or medicines that might interfere with soldiers' minds appears to be too high. The capacity of soldiers to disobey illegal and immoral orders and the duty to punish misdemeanours on the battlefield are two of the most important principles of Just War Theory. Any permissiveness in this regard would potentially open the door to a setback in the morality of warfare, which explains why a principle of precaution should always be involved when it comes to their use.

On the other hand, technologies such as exoskeletons or medicines that increase the protection of soldiers' lives and health without interfering with their moral judgement should be used without reservation. Not to use them would amount to a form of institutional negligence. While it is true that such equipment may create an asymmetrical relationship between the soldiers who possess them and those who do not, it not morally reprehensible as it does not engender a situation of riskless warfare. The same can be said with regard to the development by the US Department of Defense of a new type of military camouflage that would not only allow soldiers to hide themselves efficiently from the enemy, but also protect them against extreme heat caused by bombs (Docherty, 2012). A typical bomb blast usually emits heat of at least 600°C (or 1,000°F), which is sufficient to burn human skin. Moreover, the conventional camouflage used by soldiers only makes the effects worse as it contains chemicals that ignite more easily in such heat. The new camouflage, which is waterproof and also repels insects, can protect naked skin for a period of fifteen seconds before the appearance of first-degree burns, which gives enough time for soldiers to evacuate from the area without suffering serious harm. It is, however, true that this type of technology might at some point evolve in such a way as to make soldiers entirely invulnerable to their enemies' weaponry. If this were ever the case, the technology would challenge the moral justification of killing in warfare already discussed. However, on the basis of the current development of capacity-increasing technologies, this prospect appears to be very distant.

The main task now is to sort the wheat from the chaff and to find ways that will allow us to determine which capacity-increasing technologies should be used and which should not be, on the basis of their impact on soldiers' moral agency. This can be achieved only through a long series of experiments and testing: a process that also needs to be contained through ethical principles. This is the topic of the next chapter.

Notes

1 Principles of *jus post bellum* deal with the rules that will contribute to ending a war completely and fairly.

2 As argued in the eighteenth century by Marshal Maurice de Saxe, 'Discipline is the most important thing to instil after the troops have been formed. It is the essence of the military. If it is not established with wisdom and executed with resolute rigour, an army will be useless. The regiments and the armies would simply be a worthless armed rabble more dangerous than the enemies of the states' (Ardant du Picq, 1947, p. 39, my translation).

3 For instance, Article 83 of the Canadian National Defense Act states the following: 'Every person who disobeys a lawful command of a superior officer is guilty of an offence and on conviction is liable to imprisonment for life or to less punishment.'

4 According to the General von Clausewitz, individuals who have specific qualities, such as the ability to show initiative or to adapt quickly to the enemy's strategy, should not be subjected to a higher degree of obedience than others; on the contrary, they should have superior commanding responsibilities. This was one of the main factors that allowed Napoleon to be so successful, as he was surrounded by officers who had such qualities. Joachim Murat, André Masséna, Michel Ney, and Jean Lannes are good examples. Napoleon allowed the latter a high degree of freedom on the battlefield because he was a man of extraordinary bravery who was always calm during battles and had a penetrating eye which allowed him to benefit from enemies' mistakes. It was from this perspective that Napoleon famously once said that the best quality of a general is his capacity to disobey. Obedience is, in contrast, seen as a way to compensate for the lack of these natural qualities.

5 For his action in My Lai, he was awarded the Distinguished Flying Cross. However, it was not awarded for his actual behaviour on that fateful day, as the citation was formulated to cover up the massacre. It praised him for taking a Vietnamese child 'caught in intense crossfire' to a hospital and also stated that his 'sound judgement had greatly enhanced Vietnamese–American relations in the operational area' (Allison, 2012, p. 73).

6 In a letter to one of his friends, Major Edward W. Wynkoop, Soules wrote the following: 'I refused to fire, and swore that none but a coward would, for by this time hundreds of women and children were coming towards us, and getting on their knees for mercy. I tell you Ned it was hard to see little children on their knees have their brains beat out by men professing to be civilized. … I saw two Indians hold one of another's hands, chased until they were exhausted, when they kneeled down, and clasped each other around the neck and were both shot together. They were all scalped, and as high as half a dozen taken from one head. They were all horribly mutilated. One woman was cut open and a child taken out of her, and scalped. … Squaw's snatches were cut out for trophies. You would think it impossible for white men to butcher and mutilate human beings as they did there' (Roberts and Halaas, 2004, pp. 325–326).

7 In the Netherlands, the public shares the same opinion as that held by the court: the word 'Karremans' – which refers to the surname of the commanding officer of the Dutch battalion at Srebrenica, Thom Karremans – is often used to refer to helpless passivity or cowardice in a threatening situation.

8 As the court's decision states, 'Dutchbat [the name of the Dutch United Nations battalion] knew when the evacuation of the refugees of the compound became relevant that the men selected and carried off by the Bosnian Serbs faced death or inhumane treatment. It has therefore also been considered that by the end of the afternoon of 13 July 1995, Dutchbat must have been aware of the serious danger (*serious risk*) of genocide if the men residing at the compound were to be carried off by the Bosnian Serbs' (Rechtbank Den Haag, 2014, par. 4.324).

9 The leading role of the media in the genocide was later recognized by the International Criminal Tribunal for Rwanda, which concluded that 'The newspaper and the radio explicitly and repeatedly, in fact relentlessly, targeted the Tutsi population for destruction. Demonizing the Tutsi as having inherently evil qualities, equating the ethnic group with "the enemy" and portraying its women as seductive enemy agents, the media called for the extermination of the Tutsi ethnic group as a response to the political threat that they associated with Tutsi ethnicity' (Prosecutor v. Nahimana, Barayagwiza, and Ngeze, Case No. ICTR 99-52-A, par. 72).

10 Huntington cannot be clearer when he writes that 'The causes of war are always political. State policy aimed at continuing political objectives precedes war, determines the resort to war, dictates the nature of the war, concludes the war, and continues on after the war. War must be the instrument of political purpose' (Huntington, 1957, p. 65).

11 For Brian Imiola, this principle of non-responsibility has been respected in recent history. As he writes, 'In general, punishment has not occurred at the conclusion of wars in the nineteenth, twentieth and twenty-first centuries. Wehrmacht soldiers after the Second World War and Iraqi soldiers after the Gulf War were not viewed as guilty for the crime of war nor punished for fighting for an unjust cause' (Imiola, 2014, p. 21).

12 To humiliate the French, Hitler had them sign the act of surrender in the same railway carriage in which the German delegation had asked for an armistice in November 1918, which ultimately led to the Versailles diktat.

13 He was sentenced to four years' imprisonment and a fine of 100 francs.

14 For which he was sentenced to death *in absentia*.

15 These officers told French soldiers: 'You have the freedom to serve under the command of de Gaulle. But we must tell you that if you decide to do so, you will be considered to be rebels against your government' (de Gaulle, 1954, p. 75, my translation).

16 On 15 December 1942, Hitler ordered that three hundred thousand Germans workers be transferred into the armed forces.

17 According to this principle, weapons and means and instruments of warfare must be able to make a distinction between combatants and non-combatants. Thus 'Biological weapons and most anti-personnel landmines … are indiscriminate and therefore illegal in that they cannot distinguish whether they are about to infect or blow up a small child versus an enemy combatant' (Lin, 2013).

18 As defined by Lin, this principle 'demands that the use of a weapon be proportional to the military objective, so to keep civilian casualties to a minimum. For instance, dropping a nuclear bomb to kill a hidden sniper would be a disproportionate use of force, since other less drastic methods could have been used' (Lin, 2013).

19 For Lin, this principle also relates to proportionality 'in that it requires methods of attack to be minimally harmful in rendering a warfighter *hors de combat*, or unable to fight. This prohibition has led to the ban of such weapons as poison, exploding bullets, and blinding lasers, which cause more injury or suffering than needed to neutralize a combatant' (Lin, 2013).

20 That capacity-increasing technologies might alter soldiers' mind and transform them into killing machines is not the only fear. Indeed, we cannot ignore the fact that the prospect of cognitive changes associated with these technologies can play a counter-productive role with regard to the military's duty of care by leading it to accept what can be labelled suicidal missions. As reported by Norman Ohler, the use of Pervitin played such a role by erasing soldiers' fear of death, which led their officers to order them to carry out highly dangerous tasks. For instance, during the Battle of France in May 1940, German units under the effects of Pervitin launched repetitive suicidal assaults against the enemy

position without questioning the legitimacy of their orders. This suicidal mission was nonetheless accepted without hesitation by the doped members of the *Wehrmacht*, who took the position at high cost after it was evacuated by the Belgians, who only lost eleven men in the battle (Ohler, 2016, p. 83). Part of the courage manifested by kamikaze pilots in the closing months of the Second World War might also be attributed to the use of methamphetamine. As Lukasz Kamienski mentions, we now know that most were not volunteers at all, according to some of their diaries and the recollection of kamikaze pilots who managed to survive (Kamienski, 2016, p. 129). When the pilots were torn by fear and doubts, 'methamphetamine helped mitigate these emotions a bit. By empowering them with greater confidence and energy, the drug might also have fostered the irrational belief that they might fulfill their duty, survive, and somehow return home in glory' (Kamienski, 2016, p. 130).

21 Article 33.1 (1) of the Canadian criminal code states: 'It is not a defence to an offence that the accused, by reason of self-induced intoxication, lacked the general intent or the voluntariness required to commit the offence.'

22 Once again, as the judges of the Canadian Supreme Court have argued, 'A malfunctioning of the mind that results exclusively from self-induced intoxication cannot be considered a disease of the mind in the legal sense, since it is not a product of the individual's inherent psychological makeup' (Supreme Court of Canada, 2011).

23 A good example of this would be a driver hitting a pedestrian after his food or drink was spiked with a drug without his knowledge.

24 As Bale said, he simply took 'whatever the military gave him' (James, 2013).

25 As May writes, 'The victims of war or armed conflict often find it difficult to reach closure without a trial that prosecutes those who killed their relatives or harmed them. While it is good to aim for reconciliation, so that a society can move on, it is a mistake to do so at the expense of victims and at the further cost of allowing perpetrators to think that they can act with impunity' (May, 2012, p. 81).

26 As mentioned by the US President's Council on Bioethics, 'At first glance, such a drug would seem ideally suited for the prevention of PTSD, the complex of debilitating symptoms that sometimes afflict those who have experienced severe trauma. These symptoms – which include persistent re-experiencing of the traumatic event and avoidance of every person, place, or thing that might stimulate the horrid memory's return – can so burden mental life as to make normal everyday living extremely difficult, if not impossible. For those suffering from these disturbing symptoms, a drug that could separate a painful memory from its powerful emotional component would appear very welcome indeed' (*Beyond Therapy*, 2003, pp. 253–254).

27 This technology is integral to the 'Brain-Machine Interface' program of DARPA.

28 After interviewing soldiers who participated in the Normandy beach landings, General George Marshall learned that fatigue was responsible for an overwhelming number of casualties.

29 In 1997, the researchers also tested a similar experiment on Vietnam War veterans by using a virtual reality game labelled *Virtual Vietnam* (Halpern, 2008).

References

Allison, William Thomas. 2012. *My Lai: An American Atrocity in the Vietnam War*. Baltimore: Johns Hopkins University Press.

Ardant du Picq, Charles. 1947. *Battle Studies*. Harrisburg, Pennsylvania: Stackpole Books.

Bakker, Jeanne L. 1989. 'The Defense of Obedience to Superior Orders: The Mens Rea Requirement', *American Journal of Criminal Law*, Vol. 17, No. 1, pp. 55–80.

Bernton, Hal. 2013. Did Malarial Drug Play a Role in Bales' Afghan Murders?. *The Seattle Times*, July 18.

Beyond Therapy: Biotechnology and the Pursuit of Happiness. 2003. A Report of the President's Council on Bioethics. Washington, DC: Dana Press.

Caron, Jean-François. 2014. 'An Ethical and Judicial Framework for Mercy Killing on the Battlefield', *Journal of Military Ethics*, Vol. 13, No. 3, pp. 228–239.

Caron, Jean-François. 2017. 'Exploring the Extent of Ethical Disobedience through the Lens of the Srebrenica and Rwanda Genocides: Can Soldiers Disobey Lawful Orders?', *Critical Military Studies*, February. www.tandfonline.com/doi/abs/10.1080/23337486.2017.127892 0?journalCode=rcms20 (last accessed 4 October 2017).

City of Minneapolis, City of Minneapolis v. Altimus, 238 N.W.2d 851 (Minn. 1976).

Court of Appeals of California, People v. Scott (1983), 146 Cal. App. 3d 823, 194 Cal. Rptr. 633.

Dallaire, Roméo. 2003. *Shake Hands with the Devil: The Failure of Humanity in Rwanda*. Toronto: Random House Canada.

De Gaulle, Charles. 1954. *Mémoires de guerre: l'appel 1940–1942*. Paris: Plon.

District Court of Appeal of Florida, Carter v. State, 710 So. 2d 110 (Fla. Dist. Ct. App. 1998).

Docherty, Molly. 2012. 'Heat-Proof Face Paint to Withstand Bomb Heat', *New Scientist*, 22 August.

Foucault, Michel. 1975. *Surveiller et punir: naissance de la prison*. Paris: Gallimard.

Gandhi, Mahatma. 1925. *Young India*, Vol. 12, No. 3. 1925, pp. 88–89.

Girardet, Raoul. 1960. 'Pouvoir civil et pouvoir militaire dans la France contemporaine', *Revue française de science politique*, No. 1, pp. 5–38.

Girardet, Raoul. 1999. 'La désobéissance légitime 1940–1962', in Olivier Fourcade, Éric Duhamel, and Philippe Vial (eds), *Militaires en République 1870–1962: les officiers, le pouvoir et la vie publique en France*. Paris: Sorbonne, pp. 547–552.

Green, Leslie Claude. 1989. 'Superior Orders and Command Responsibility', *Canadian Yearbook of International Law*, Vol. 27, pp. 167–202.

Halpern, Sue. 2008. 'Virtual Iraq. Using Simulation to Treat a New Generation of Traumatized Veterans', *New Yorker*, 19 May.

Homeland Security News Wire. 2010. 'In the Trenches: New Helmets to Make Soldiers More Alert, Reduce Stress, Pain'. www.homelandsecuritynewswire.com/new-helmets-make-soldiers-more-alert-reduce-stress-pain (last accessed 4 October 2017).

Huntington, Samuel P. 1957. *The Soldier and the State: The Theory and Politics of Civil–Military Relations*. Cambridge, Massachusetts: Harvard University Press.

Imiola, Brian. 2014. 'The Duty of Diligence: Knowledge, Responsibility, and Selective Conscientious Objection'. In Andrea Ellner, Paul Robinson, and David Whetham (eds), *When Soldiers Say No: Selective Conscientious Objection in the Modern Military*. New York: Routledge, pp. 19–39.

James, Susan Donaldson. 2013. 'Antimalarial Drug Linked to Sgt. Robert Bales Massacre', ABC News, 22 July. http://abcnews.go.com/Health/antimalarial-drug-linked-sgt-robert-bales-massacre/story?id=19713961 (last accessed 4 October 2017).

Kamienski, Lukasz. 2016. *Shooting Up: A Short History of Drugs and War*. Oxford: Oxford University Press.

Ketchum, James S. 2012. *Chemical Warfare Secrets Almost Forgotten: A Personal Story of Medical Testing of Army Volunteers during the Cold War (1955–1975)*. Bloomington, Indiana: Author House.

Khatchadourian, Raffi. 2012. 'Operation Delirium', *New Yorker*, 17 December.

Leveringhaus, Alex. 2015. 'Assigning Responsibility in Enhanced Warfare'. In Jai Galliott and Mianna Lotz (eds), *Super Soldiers: The Ethical, Legal and Social Implications*. Farnham: Ashgate, pp. 141–152.

Lin, Patrick. 2013. 'Could Human Enhancement Turn Soldiers into Weapons that Violate International Law? Yes', *New Atlantic*, 4 January.

May, Larry. 2012. *After War Ends. A Philosophical Perspective*. Cambridge: Cambridge University Press.

McCullough, David. 1992. *Brave Companions: Portraits in History*. New York: Simon & Schuster, p. 132.

Montgomery, Nancy. 2005. 'Maynulet Testifies in Own Defense, Says Killing Wounded Iraqi Right Thing to Do', *Stars and Stripes*, 31 March.

Moore, Thomas J., Joseph Glenmullen, and Curt D. Furberg. 2010. 'Prescription Drugs Associated with Reports of Violence towards Others', *PLOS One*, 15 December. http://journals.plos.org/plosone/article?id=10.1371/journal.pone.0015337 (last accessed 4 October 2017).

Mount, Mike and Elain Quijano. 2009. 'Iraqi Insurgents Hacked Predator Drone Feeds, U.S. Official Indicates', CNN, 18 December. http://edition.cnn.com/2009/US/12/17/drone.video.hacked/index.html (last accessed 4 October 2017).

National Defence and the Canadian Armed Forces. 2016. 'Post-Traumatic Stress Disorder', 13 January. www.forces.gc.ca/en/news/article.page?doc=post-traumatic-stress-disorder/hjlbrhp4 (last accessed 4 October 2017).

Ohler, Norman. 2016. *L'extase totale: le IIIe Reich, les Allemands et la drogue*. Paris: La Découverte.

Prosecutor v. Nahimana, Barayagwiza, and Ngeze. Case No. ICTR 99-52-A, International Criminal Tribunal for Rwanda, Appeals Chamber, 28 November 2007.

Quebec Court, R. c. Côté (2011 QCCQ 8227).

R. v. Finta, [1994] 1 SCR 701.

Rechtbank Den Haag. 2014. ECLI:NL:RBDHA:2014:8748, 17 July. http://uitspraken.rechtspraak.nl/inziendocument?id=ECLI:NL:RBDHA:2014:8748 (last accessed 4 October 2017).

Roberts, Gary L. and David Fridtjof Halaas. 2004. Written in Blood: The Soule-Cramer Sand Creek Massacre Letters'. In Steve Grinstead and Ben Fogelberg (eds), *Western Voices: 125 Years of Colorado Writing*. Golden, Colorado: Fulcrum, pp. 320–337.

Shapiro, Nina. 2013. 'Mefloquine Monday', *Seattle Weekly News*, 24 July. http://archive.seattleweekly.com/news/947794-129/mefloquine-bales-says-drug-nevin-alderate (last accessed 4 October 2017).

Supreme Court of Canada. 2011. R. v. Bouchard-Lebrun, 2011 CSC 58, [2011] 3 RCS 575.

Thompson, Allan (ed.). 2007. *The Medias and the Rwanda Genocide*. Ann Arbor: Pluto Press, pp. 1–12.

Uniform Code of Military Justice, 64 Stat. 109, 10 USC §§ 801–946. Décret n°75-675 du 28 juillet 1975 portant règlement de discipline générale dans les armées.

United States v. Maynulet, M.J. 2009. No. 09-0073/AR (CAAF 1 June 12009).

US Department of Veterans Affairs. n.d. 'How Common is PTSD?'. https://www.ptsd.va.gov/public/ptsd-overview/basics/how-common-is-ptsd.asp (last accessed 4 October 2017).

Vattel, Emer de. 1916. *The Law of Nations, or the Principles of Natural Law.* Washington, DC: Carnegie Institution.

Von Clausewitz, Carl. 1976. *On War.* Oxford: Oxford University Press.

Walzer, Michael. 1970. *Obligations: Essays on Disobedience, War, and Citizenship.* Cambridge, Massachusetts: Harvard University Press.

Wenker, Kenneth H. 1981. 'Morality and Military Obedience', *Air University Review*, Vol. 32, No. 5 (July–August), pp. 76–83. http://www.au.af.mil/au/afri/aspj/airchronicles/aureview/1981/jul-aug/wenker.htm (last accessed 4 October 2017)

White, Stephen E. 2008. 'Brave New World: Neurowarfare and the Limits of International Humanitarian Law', *Cornell International Law Journal*, Vol. 41, No. 1, pp. 177–210.

Wolfendale, Jessica. 2008. 'Performance-Enhancing Technologies and Moral Responsibility in the Military', *American Journal of Bioethics*, Vol. 8, No. 2, pp. 28–38.

4

The ethics of developing
capacity-increasing technologies

As discussed in the previous chapter, there is a need to establish criteria in order to determine which capacity-increasing technologies can be used on the battlefield without infringing moral principles. However, this debate would not be complete without a reflection on how these technologies ought to be developed. More precisely, we need to analyse how the technologies should be tested on soldiers and how the members of the military should comply with their use. From this perspective, a discussion of the ethics of military research is coextensive with the issue of the super soldier. Moreover, in view of the manner in which soldiers have been treated by the military in the not so distant past, such a discussion is all the more necessary in order to avoid future disregard for their health. Indeed, as has already been stated, combatants have very often been used as guinea pigs by their respective states or militaries. In view of the nature of what seems to be occurring in the military, many people are understandably fearful that current and future capacity-increasing technologies might follow the same path. This is largely because history is filled with awful stories of soldiers who have been used as expendable goods in the name of science, contravening the institutional duty of care the military owes to its members.

This occurred within totalitarian regimes of the past. In view of the inhumane nature of the Nazis, few would be surprised by the fact that German soldiers were treated simply as a means to a higher end, especially in the case of the drug Pervitin, mentioned above. This was initially created to treat depression, asthma, and blood flow problems, but the German army quickly realized its potential. Otto Friedrich Ranke, a military doctor and head of the Institute of Physiology and Defence at the Berlin Military Academy of Medicine, noticed how it was able to increase people's capacity to concentrate for long hours and to ignore pain and fatigue, and could keep someone awake for a whole night more cheaply than coffee (four pills of Pervitin cost 16 pfennig, as against 50 pfennig for fifty grams of coffee). He recommended the drug to the German High Command as part of the standard equipment of deployed soldiers, despite recognizing its risks to their health. The sacrifice was high: many soldiers became addicted after only two or three weeks of consumption and others, who were suffering from cardiac problems, died after mixing it with alcohol. However, its use was never really restricted. For the German

High Command, military necessities were a legitimate end that justified sacrificing the members of the *Wehrmacht*.

We might also recall the numerous nuclear tests carried out by the Soviet Union, which exposed thousands of soldiers to radiation. The most famous is undoubtedly the Totskoye exercise, undertaken in 1954, which saw an army of forty-five thousand soldiers march through the epicentre of a nuclear blast of 28 kilotons of TNT in order to determine whether troops could fight a battle in an area immediately after it was hit by an atomic bomb. After the collapse of the Soviet Union, documents obtained from the secret military archives stated that these men were exposed to high radioactivity for an extended period of time without appropriate protective gear. At the time this information was disclosed, about a thousand of the former Soviet soldiers who took part in the test were still alive; according to them, they never received adequate treatment for the various health problems they suffered in the aftermath of their exposure (Simons, 1993).

However, complete disregard of soldiers' health is not solely confined to inhumane or totalitarian regimes. For example, the United States deliberately exposed thousands of soldiers to radiation during the various atmospheric nuclear tests it conducted, such as Operations Plumbbob, Castle Bravo, or Crossroads.[1] The Soviets did the same in 1954 during the Totskoye nuclear exercise, where forty-five thousand soldiers were exposed to an atomic explosion twice as powerful as the one in Hiroshima. The consequences for the US servicemen who were exposed were devastating for their health. For instance, reports have shown that participants in nuclear experiments have an increased risk of suffering from diseases of the digestive system, particularly cirrhosis of the liver (Raman et al., 1987), cancer of the digestive organs (Watanabe, Kang, and Dalager, 1995), and haematological cancer, particularly leukaemia (Pearce et al., 1990; Pierce et al., 1996; Caldwell et al., 1983). Another example is Project Shipboard Hazard and Defense, conducted in the 1960s by the US Army. This was a large effort to analyse how sailors aboard military ships could detect and respond to chemical and biological attacks while maintaining their fighting capacities. On forty-six occasions, the servicemen were exposed, without their consent, to numerous substances such as the nerve gases VX and sarin and other toxic chemicals, as well as a variety of biological agents (Blum, 2006, pp. 152–154). During the Second World War, the US Navy also tested sulphur mustard and Lewisite, poisonous gases used in chemical weapons, on 2,500 of its sailors., in an exercise known as Operation Man Break. These men were given gas masks and special clothes to wear and were escorted into a chamber, which was then locked from the outside, after which the gases were released inside. However, a seventeen-year-old navy seaman named Nathan Schnurman experienced problems with his mask and asked permission to come out: a demand that was refused. He then vomited into his mask, passed out, and had a heart attack. When he awoke, he had painful blisters on most of his body. He was given no medical treatment and was ordered never to speak about what he experienced, under the threat of being tried for treason. When other sailors became afraid to participate in the experiment, they received threats that left them with no choice. As Seaman Russell O'Berry, then a seventeen-year-old Virginian, stated, 'Some of the men refused to go into the gas chamber and were given a direct order. We were told if we did not go through with it we would get 40 years in Fort Leavensworth in Kansas (the army prison)' (Cockburn, 1993).

Great Britain also famously tested a highly dangerous nerve gas on more than 3,400 subjects between 1945 and 1989 at Porton Down, a place once described by a Whitehall official as 'a sinister and nefarious establishment' (Evans, 2004). Test subjects were exposed to high doses of sarin in order to determine how this chemical would impair the mental performance and intellectual capacities of soldiers. The repercussions of the gas were terrible. Between 1952 and 1953, six soldiers were hospitalized as a result of exposure to nerve agents, and another – a leading aircraftsman, Ronald Maddison – died on 6 May 1953 after his forearm came into contact with 200 mg of pure sarin (Schmidt, 2006). In 2004, the High Court ruled that he was unlawfully killed.

On the basis of these few examples, it is clear that the military has not hesitated to overrule ethical norms in this type of research. This is why many feel that the military-industrial complex is untrustworthy when it comes to the treatment of soldiers as individuals, often treating them simply as a means to an end. It is astonishing to note that animals are very often treated with more humanity than test subjects in the military (Festing and Wilkinson, 2007). There is an undeniable need to protect soldiers from these experiments and to determine ways in which military research may be supervised.

Although the civilian bioethics principles established in the aftermath of the Second World War are sound and able to protect individuals efficiently, some authors have argued that they have little application in the context of the military because of the fundamental differences between civilian and military values (Mehlman and Corley, 2014). While this is true, the ethical solutions currently being put forward are unsatisfactory as they still tend to go against Kantian ethics. Hence there is a need to establish strong ethical guidelines for military research that are not only adapted to the reality of the battlefield, but also respectful of Kant's moral philosophy. That is what this chapter will try to establish.

The discussion will be divided into two parts. The first will highlight the main components of civilian bioethical principles and present the problems they apparently pose for military research. More specifically, it will analyse how the notion of consent, which is so fundamental in civilian research, has to be significantly altered for soldiers by various factors that are peculiar to the world of the military. This conflict is undoubtedly the heart of the problem. On the basis of these considerations, the second part will attempt to determine ethical guidelines for the development of capacity-increasing technologies and the ways they should be used by soldiers.

The value of civilian bioethics principles and their applicability to problems for the military

Following the Second World War and the discovery of the savagery of the Nazi regime, the international community established the Nuremberg Code on 20 August 1947, in order to determine a set of research principles for human experimentation.

This was necessitated by the fact that the twenty-three doctors on trial, for forced sterilization of more than three million people, conducting abortions on women deemed genetically deficient, freezing experiments,[2] high-altitude experiments,[3] malaria experiments,[4] mustard gas testings,[5] and euthanasia, argued that their experiments were not very different from pre-war experiments and that they could not be considered illegal, as there was no legislation that differentiated legal experiments from illegal ones. The doctors' trial also showed that these terrible acts were mainly motivated by the idea that the purity and general health of German society were more important than the health and life of its individual members. Because of this reasoning and the lack of distinction between morally acceptable and unacceptable experiments, the Nuremberg Code was implemented with its now famous ten requirements, which serve today as general guidelines for medical research. These points are as follows:

1. The voluntary consent of the human subject.
2. The experiment should be such as to yield fruitful results for the good of society, unprocurable by other methods or means of study, and not random and unnecessary in nature.
3. The experiment should be so designed and based on the results of animal experimentation and a knowledge of the natural history of the disease or other problem under study that the anticipated results will justify the performance of the experiment.
4. The experiment should be so conducted as to avoid all unnecessary physical and mental suffering and injury.
5. No experiment should be conducted where there is an *a priori* reason to believe that death or disabling injury will occur, except, perhaps, in those experiments where the experimental physicians also serve as subjects.
6. The degree of risk to be taken should never exceed that determined by the humanitarian importance of the problem to be solved by the experiment.
7. Proper preparations should be made and adequate facilities provided to protect the experimental subject against even remote possibilities of injury, disability, or death.
8. The experiment should be conducted only by scientifically qualified persons. The highest degree of skill and care should be required through all stages of the experiment of those who conduct or engage in the experiment.
9. During the course of the experiment, the human subject should be at liberty to bring the experiment to an end if he has reached the physical or mental state where continuation of the experiment seems to him to be impossible.
10. During the course of the experiment the scientist in charge must be prepared to terminate the experiment at any stage, if he has probably cause to believe, in the exercise of the good faith, superior skill and careful judgement required of him that a continuation of the experiment is likely to result in injury, disability, or death to the experimental subject. (*Trials of War Criminals before the Nuremberg Military Tribunals under Control Council Law*, 1949, No. 10, Vol. 2, pp. 181–182)

These principles were reiterated in 1964, and others were also added, in what is now called the Declaration of Helsinki. The basic principles stated in the document outline the duty of the physician to promote and safeguard the health, well-being, and rights of patients, including those who are involved in medical research; the basic idea is that while the primary purpose of medical research is to generate new knowledge, this goal should never take precedence over the rights and interests of individual research subjects. In medical research involving human subjects capable of giving informed consent, each potential subject must be adequately informed of the aims, methods, sources of funding, possible conflicts of interest, institutional affiliations of the researcher, and anticipated benefits and potential risks of the study, as well as the discomfort it may entail, the post-study provisions, and any other relevant aspects.

However, it is clear from the numerous experiments performed in the aftermath of the Second World War that this code was never fully respected. For instance, it took until 1953 for the Pentagon to specify the obligation of written consent from soldiers participating in military research. When the Edgewood Arsenal experiments became public in the 1970s, it was clear that they had not complied with the policy at all (Moreno, 2012, p. 44).

For certain authors, these principles are scarcely suitable for application to military experiments because the core military values are fundamentally antithetical to civilian research. For example, Anthony E. Hartle argues that since the purpose of the military institution is to protect the nation from domestic and foreign enemies, 'members of the profession subordinate personal welfare to [its] welfare' (Hartle, 2004, pp. 59–60). For its part, the US Supreme Court stated in 1974 that 'Within the military community, there is simply not the same [individual] autonomy as there is in the larger civilian community' (Parker v. Levy, 417 US 733, 743 (1974)).

It has to be admitted that soldiers– if they are not conscripts – have indeed agreed voluntarily to relinquish certain rights and privileges by joining the military, such as their capacity to remain healthy and alive. Indeed, unlike civilians who go to work on a daily basis, soldiers may be asked to risk their lives during dangerous or routine missions. Moreover, when they are given a lawful order, they do not have the opportunity to discuss or negotiate its purpose or method with their superior, as is very often the case for civilians. They must simply obey, even if they feel that the order is unreasonable or that the goal of the mission could be achieved in a more effective way.[6] However, as mentioned in the second chapter, the military institution has to fulfil its duty of care towards its members. As a consequence, it must not treat them as expendable goods or without regard for their health and safety. Accordingly, exposing them to radiation, nerve gas, or bacteriological weapons as mere guinea pigs is contrary to this moral obligation and cannot be justified even for the sake of the common good of the nation they are serving.

One main incompatibility between the civilian and the military reality undoubt- edly concerns the question of consent. In the civilian situation, individuals involved in research should have the legal capacity to give consent and be able to exercise free power of choice without the intervention of any element of force, fraud, deceit, duress, over-reaching, or other ulterior forms of constraint or coercion, as well as receive sufficient details and explanation of the subject matter involved to enable them to make an enlightened decision. This latter element requires that before he or she agrees to participate the experimental subject should be told the nature, duration, and purpose of the experiment; the method and means by which it is to be conducted; all inconveniences and hazards reasonably to be expected; and the effects upon her or his health or person which may possibly come from participation in the experiment. However, this informed consent is not required in the military as it is an organization that can interfere with its members' wishes and preferences in order to protect them and their colleagues from potential harm. This paternalism explains why soldiers are sometimes forced to take particular medicines. This situation arose within the US Army during the First Gulf War in 1991. Fearing that Saddam Hussein's army might use nerve agents, the Department of Defense ordered that all the soldiers deployed in the Persian Gulf should be given pyridostigmine bromide (BP) and botulinum toxin (BT) without their consent, even if they were investigational drugs.[7] Fearing that this might be illegal under the Nuremberg Code, the Department of Defense asked for permission to establish a special procedure, known as an 'interim rule', which would it allow it to give these products without the soldiers' consent. This was granted by the Food and Drug Administration (FDA) for the administration of investigational drugs for military purposes. Some Gulf War veterans later challenged this rule, but all were dismissed by the courts (Doe and Doe v. Sullivan, 756 F. Supp. 12 (USDC 1991); Doe v. Sullivan, 938 F.2d 1370 (US App. DC 1991)). In 1999, the right to waive soldiers' consent for these types of drugs was transferred directly to the President (Executive Order 13139). In 2004, the US Congress gave elected officials another way to waive soldiers' consent by implementing the Emergency Use Authorization, which allows the use of unapproved drugs or vaccines in case of a national emergency. Similarly, in 1998, the US military decided that all its members should be vaccinated against anthrax. At the time, the vaccine was not approved by the FDA for airborne exposure, but it was nonetheless given to soldiers without their consent.

Such examples show a clear tension between the duty of care of the military and the possible problems it could lead to. We can presume that if faced with a potential life-threatening bacteriological warfare, the military would want to deploy all the means at its disposal – even untested ones – to ensure the best protection for its members. However, at the same time, this paternalism might simply backfire, as the administration of untested medicines or drugs can lead to unexpected health issues. Although this has never been proved, it is impossible to ignore the fact that Gulf War syndrome, a chronic multi-symptom disorder which affected an estimated two hundred and fifty thousand of the seven hundred thousand soldiers deployed in the 1991 Gulf War, is associated with various vaccines as a possible cause (McManus et al., 2005, p. 1124). Moreover, as has been discussed in the previous chapters, the administration of drugs and medicines without soldiers' consent could open the door to cases of involuntary intoxication, which would deprive members of the armed forces of criminal responsibility and run counter to the principles of *jus post bellum*. This begs the question of what should be done: an obvious question, but one whose answer is far less evident.

Although the application of civilian research ethics might seem appealing as a means to avoid these issues, it must be acknowledged that such a possibility raises a significant problem. Indeed, there is a need to consider the potential negative consequences of this principle of consent.[8] Various arguments can be raised against the proposal. First of all, the need to obtain soldiers' consent for every medicine they might have to take may simply be impractical. For instance, some missions require urgency and rapid deployment, such as catching a terrorist leader who has a reputation of changing his location on a daily basis. We can easily imagine that performing such operations might require the use of certain drugs to increase soldiers' chances of survival. However, allowing them the right to obtain a full disclosure of the associated potential risks – which implies granting them sufficient time to discuss these with doctors who have knowledge of them – might simply delay the mission and prevent the military from achieving what could be a legitimate goal.

But if we leave this strategic reason aside, there is also a moral argument in favour of not granting soldiers the right to refuse to use certain drugs. More precisely, soldiers' right to fulfil their duties with limited impact on their health and safety does not depend exclusively on the military: it also depends on their comrades, because individuals serving in the same unit are interdependent. A mistake made by a soldier might increase the risk of death for her or his fellow soldiers. This is the reason why a sentinel found asleep or drunk at her or his post is usually punished severely. This was the case for Private Jack Dunn during the Battle of Gallipoli in 1915. after which he was sentenced to death by a court martial.[9] Therefore soldiers are also an element of the duty of care and, consequently, have a collective obligation towards others.

This duty has implications for the use of technologies and medicines, because if soldier refuse to use them, not only do they increase their own vulnerability, but that exposure may also make their comrades more likely to be harmed. In other words, the fear of protecting oneself from the potential ill-effects of drugs and vaccines in the military may actually lead to situations where members of a unit have to proceed to a dangerous evacuation on the battlefield, thereby increasing their exposure to death. Bearing in mind the camaraderie and the ties that unite soldiers (for which they have often referred to each other as 'brothers in arms'), we can assume that they might still agree to use capacity-increasing technologies despite their potential risks. As argued by Michael Frisina,

> The success of small unit tactics depends upon unit members being able to perform their assigned tasks when called upon to do so. Unprotected soldiers suffering injury from chemical and biological agents become liabilities to the welfare of their unit members when they are unable to perform their assigned role. Ultimately, the success of the overall mission is potentially jeopardized. If there is a derived benefit from taking these investigational agents and some members fail to accept this benefit, the negative effects of biological and chemical weapons not only impacts on those members not protected but degrades the capability of the entire unit and ultimately the welfare of the other unit members. (Frisina, 2003, p. 551)

The same argument has been used by Patrick Lin, Maxwell Mehlman, and Keith Abney, for whom 'The flip-side of consent is refusal to consent, and warfighters are likely to be reluctant to refuse to use a performance enhancement if they think that this would adversely affect other members of the unit, for example, by placing on them more risk or more of the responsibility for carrying out the mission' (Lin, Mehlman, and Abney, 2013, p. 74). This risk is an element of the military's paternalistic nature and a reason why its members' personal preferences may be legitimately suppressed (Wolfendale and Clarke, 2008, p. 341). In order to avoid the collateral harm that could result from the need to obtain soldiers' consent before using medicines or other technologies, it is therefore necessary for the military institution to show a form of paternalism by subordinating the individual's interests to the collective welfare.

In view of this, how is it possible to create a situation that would allow soldiers to waive consent without exposing them to risks to their health and moral responsibilities? Needless to say, this balance is very difficult to establish. For instance, the Presidential Advisory Committee on Gulf War Veterans' Illnesses (1997) and the Advisory Committee on Human Radiation Experiments (1995) have suggested that, given the voluntary nature of enlistment in the military, potential recruits should be warned that they may receive vaccines or have to take medicines during their service and that joining the military will be treated as a willingness on their part to accept them. Maxwell J. Mehlman and Stephanie Corley have proposed an interesting framework. For them, respect for the welfare of soldiers can be achieved if the military observes certain principles, namely, proportionality and paternalism (Mehlman and Coley, 2014). In practice, this means that military officials should analyse whether 'the risks and benefits of [vaccines, drugs or medicines] [are] understood as well as possible and that imposing them on troops [is] necessary in order to accomplish a legitimate military objective, in that there is no less risky alternative to achieve the mission' (Mehlman and Corley, 2014, p. 342). Moreover, people who are ultimately responsible for taking the decision should act only 'on the basis of the best evidence of safety and efficacy available within the existing timeframe' (Mehlman and Corley, 2014, p. 342).

Both these frameworks, however, suffer from significant flaws. The theory of 'anticipated consent' is too generous. As Mehlman and Corley have argued,

> In the first place, describing the number and types of biomedical risks that enlistees might encounter with any specificity would be cumbersome and probably unintelligible; the alternative of describing them generally (e.g. 'you could be asked to serve in risky medical experiments or given experimental agents without your consent') is unlikely to provide sufficient knowledge of risks and benefits to permit truly informed consent to take place. In addition, it may not be possible to anticipate future types of risks at the time of enlistment. Finally, warfighters still need the protections afforded by paternalism and proportionality so that they are not deemed to have consented at the time of enlistment to disproportionate biomedical risks. (Mehlman and Corley, 2014, p. 339)

On the other hand, despite the inherent value of Mehlman and Corley's thesis, it remains problematic and dependent on an important variable: the willingness of the military or the government to provide adequate information regarding a potential threat. We have been accustomed in recent years to exaggerations on the part governments and the military, the most famous case being the invasion of Iraq in 2003, which was largely justified by the supposed possession by Saddam Hussein of various types of weapons of mass destruction. We now know from the Chilcot Report that these assumptions were made on the basis of flawed intelligence and assessments. The same exaggerations were also linked to the desire of the US Department of Defense to administer, as a matter of national emergency, anthrax vaccine to its soldiers. Many concerns have been raised that, at the time when it announced its decision, the US government did not have sufficient data on the safety and effectiveness of the vaccine (Committee on Government Reform, 2000, p. 3; Sidel and Levy, 2003, p. 299), that the threat was largely exaggerated (Committee on Government Reform, 2000, p. 2), and that it did not know the possible long-term effects of the vaccine on individuals (Sidel and Levy, 2003, p. 300). This raises the question of whether individuals can legitimately trust allegations put forward by governments or the military in order to force combatants to potentially expose their health and life for the sake of their nation. In fact, this has led some to argue that this procedure was in fact a form of medical experimentation and used soldiers as mere guinea pigs (Milner, 1997, p. 228). Needless to say, the use of experimental drugs during the 1991 Gulf War and during the previous decades at Edgewood and at Porton Down raises serious doubts about the sincerity of the military's historical and current desire to uphold its duty of care towards its members by treating them as ends in themselves and not simply as expendable means. If the future treatment of soldiers is similar to what it has been in the past, the traditional paternalism of the military institution is far from being a sufficient guarantee of their safety.

Moreover, it must be noted that the question of soldiers' autonomy and consent in the use of medicines of all sorts can be waived not only directly, but also indirectly by implicit pressures. First, unlike civilians, soldiers are trained to obey orders and to defend their country. This constitutes an implicit incentive on their part to accept situations that many civilians would not accept. As Victor Sidel and Barry Levy argue, the nature of the military organization will inevitably interfere with the notion of consent. 'Because they cannot simply "quit their jobs", file a grievance with a union, government agency, or professional organization, military personnel may not believe that they can truly refuse to participate in these experiments. They may feel more like a "captive audience" than like volunteers' (Sidel and Levy, 2003, p. 297). This fear has been echoed by the US House Committee on Government Reform, which has concluded that 'in a culture based on a chain of command and the power to compel, attempts at persuasion and education often devolve into intimidation' (Committee on Government Reform, 2000, p. 46).

It was this inherently hierarchical nature of the military that allowed commanders to order men to attend sessions at Edgewood Arsenal (Khatchadourian, 2012). Some soldiers may have agreed to participate unwillingly, but others felt that it was simply their duty to do so. As stated by Private Tim Josephs, who was a patient at Edgewood, 'I really felt a duty to my country to go and serve. You believed in your government. And you just wouldn't think they would give you something that would harm you intentionally' (Martin, 2012). As a former scientist at Porton Down once explained, 'If you advertised for people to suffer agony you would not get them' (*Report of a Court of Inquiry*, 1953, p. 86). This belief, which serves as an indirect incentive for soldiers, can of course prevent the military organization from respecting its duty of care. As long as there is an element of trust between the soldiers and the military, the former will be inclined to obey their superiors, which is not the case for civilians.

It is also clear that soldiers' refusal to obey often leads to negative consequences for their careers and even their lives, as was threatened to Private Josephs. Josephs mentioned that he started to have second thoughts after his arrival at Edgewood and was taken aside by an officer who told him, 'You volunteered for this. You're going to do it. If you don't, you're going to jail. You're going to Vietnam either way – before or after' (Martin, 2012). These threats undoubtedly had a major influence on soldiers' decision to participate freely in the experiments performed at Edgewood. Of course, such explicit threats are unacceptable. This is why the US Army has imposed rules that are supposed to eliminate pressure from superiors.[10] Despite this, we can nonetheless assume that there is a difference between the theory and the reality in the armed forces. Even with the best intentions, the army will remain a coercive institution that encourages conformity, loyalty, and above all obedience. The policy regarding the use of go-pills provides a good example. Even though the US Air Force insists that the use of these amphetamines is voluntary and that pilots must sign a consent form before using them,[11] this form states that pilots can be grounded if they decline. Anyone who knows the military system even remotely is aware that grounding – whatever the circumstances – can have serious implications for a pilot's career. This this policy tends to put indirect pressure on a pilot to take the pills, even if he or she technically has the option to refuse (Bower, 2003).

Finally, soldiers' consent may also be altered by another fundamental factor which is not inherent to civilian research. Just like other human beings, soldiers are led by a strong instinct for self-preservation, and we can assume that many of them will blindly take advantage of any means that will allow them to improve their immediate chances of survival on the battlefield. This point is highlighted by the experience of the former Luftwaffe bomber pilot Horst Freiherr von Luttitz, who said in an interview, 'Of course you don't abstain from Pervitin because it "might" have minor adverse effects on your health. Not when you could die at any moment anyway'.[12] Decades after the Second World War, the same logic is one of the reasons why American pilots still use go-pills during some of their flights. Lukasz Kamienski writes:

> During extremely long sorties, which at times continue for more than forty hours, pilots have no choice, despite officially having the right to choose freely. They are aware that if they do not want to fall asleep or make an error due to enormous operational fatigue, in other words, if they want to return to their base safely, then in specific circumstances they should – for their own sake – resort to stimulants. (Kamienski, 2016, p. 270)

Thus in a situation where the military encourages its soldiers (sometimes through the fear of punishment) to take medicines as the best way to increase their chances of survival on the battlefield or to avoid getting diseases, we can presume that the notion of 'free and enlightened consent' is significantly altered. We may wonder how these forms of pressure affect their criminal responsibility if these medicines and other capacity-increasing technologies lead them to commit crimes or deadly mistakes. Can we believe that soldiers using capacity-increasing technologies because of these implicit pressures are entirely morally responsible for the eventual collateral damages associated with their use? Even though they are offered the choice consent or not consent to their use, the case could be made that this choice is not entirely free. A good example of this is the use of go-pills in the US Air Force. These amphetamines may also decrease soldiers' moral judgement and increase their sense of obedience, but that is not their primary purpose. The use of these amphetamines is strictly a matter of military effectiveness, as a way to help pilots fulfil their mission and to increase their chances of survival by keeping sure them alert for the whole duration of their flight. Since fighting a war is not a nine-to-five job, soldiers must remain alert for unusually long periods of time as a matter of personal survival, and also for the comrades whom they are protecting from the sky. For instance, in Afghanistan, the average US soldier in combat gets only four hours of rest a day; sleep deprivation is the single largest factor in reducing combat performance. Not only are tired soldiers less physically able to fight and run than well-rested ones, but they also make more mistakes when operating complex weapons systems at their disposal – mistakes that can prove deadly to themselves and their comrades. What is seen as a necessity by the military is probably best exemplified in a comment made by Colonel Peter Demitry, Chief of the Air Force Surgeon General's Science and Technology division, who said, 'When a civilian gets tired, the appropriate strategy is to land, then sleep. In combat operations when you're strapped to an ejection seat, you don't have the luxury to pull over' (Hart, 2003).

However, the use of this type of medicine is not without unintended risks, as the Tarnak Farm incident which occurred near Kandahar in 2002 has shown. On that fateful night, an American F-16 dropped a 500-pound bomb on some Canadians who were conducting a night firing exercise, causing the deaths of four soldiers. The investigation found that the pilot had violated established procedures, and his letter of reprimand stated that he had flagrantly disregarded a direct order, exercised a total lack of basic flight discipline, and blatantly ignored the applicable rules of engagement. In this case, go-pills were directly mentioned as a contributing factor. It was reported that the pilot, who was ending a twenty-hour working day, took a 10 mg dose of dextroamphetamine two hours before the incident, and his defence attorney stated that the Air Force had pressured him to take the pill, which, he argued, may have impaired his judgement (Simpson, 2003).

Although the hearings that followed this tragic incident did not attribute the friendly fire to amphetamines, the pilot nonetheless blamed the incident on the use of these drugs, which have side effects such as confusion, delusions, and auditory hallucinations. We will never know whether the amphetamines were central factors of this incident. For the sake of argument, let us assume that the go-pills did indeed affect the pilot's judgement. In such a scenario, how should we determine his responsibility for the deaths of the four Canadian soldiers? On the one hand, it would be possible to argue that it is absolute since he had the choice to refuse to use the stimulants, but there are reasons to believe that he was pressured to take them. This begs the question of whether such a pressured soldier should be sanctioned or not. Would it be fair to do so? What could be the possible repercussions of an amnesty on the aforementioned principles of *jus post bellum*?

This discussion shows the difficulties of simply copying the principles of civilian research in military research. Not only is the soldier's capacity to consent to experiments or the use of capacity-increasing technologies affected directly or indirectly, but allowing the option to refuse their use may also result in exposing comrades to danger. In other words, the freedom to choose could potentially soldiers into life-threatening liabilities for others. Consequently, the military institution – which is ultimately responsible for the duty of care it owes to its members – would have solid grounds to reject soldiers' right to consent to the use of these technologies. However, in view of the sad past associated with military research in numerous countries, there are valid reasons to doubt the military's commitment to a sincere form of paternalism. As mentioned previously, this hierarchical relationship between soldiers and the military has in the past been harmful to the notion of paternalism that the latter should exert with regard to the former: a perverted dynamic that is becoming increasingly well known among soldiers, who are showing an increased scepticism towards the military's claim that it is acting for their own good. For instance, when the US Congress studied the Department of Defense's desire to implement a mandatory anthrax vaccine programme, its members concluded that many enlisted service personnel had concerns about the possible side effects of the vaccine, which led to a problem of retention in reserve units. The report noted that half of the men in an Air National Guard unit decided to resign in order to avoid receiving this vaccine (Committee on Government Reform, 2000, p. 46). Accordingly, it seems that the military needs to affirm and project a renewed and stronger version of paternalism when it comes to the use of the new devices or medicines that are at the core of capacity-increasing technologies. There is a need to think of a totally different framework which would both respect the military's duty of care and, at the same time, ensure that soldiers are not treated simply as a means, but rather as ends in themselves. This is what the next section explores.

A theoretical framework for capacity-increasing research in the military

From a strictly moral standpoint, the development and use of capacity-increasing technologies can be justified from two perspectives: as a component of the military's duty of care towards its members and as a way to increase the morality of warfare, as mentioned previously. Both these elements strongly militate in favour of their use. Needless to say, these technologies should not be employed on the battlefield if they can deprive soldiers of the capacity to disobey unlawful and immoral orders or if they may lead to situations where soldiers cannot be held responsible for their actions. This implies that military research needs to find ways to allow a distinction between technologies and medicines that do not harm these principles and those that do. This can be achieved only through a thorough research and development process that respects participants' health and safety: something that is currently lacking in the military.

It has already been established that requiring soldiers' consent is not an optimal solution, as it can ultimately lead to soldiers being harmed or killed while trying to rescue comrades who may have chosen not to use such technologies or medicines. The military would not fulfil its duty of care towards soldiers by giving this kind of freedom. However, if consent is waived for members of the armed forces, this implies that they have full confidence that the military will not order them to use things that might be harmful to their health. In other words, the fear that their institution might betray their trust by ordering them to use harmful technologies or medicines must be eliminated. Needless to say, this paternalism has been hampered by numerous relatively recent examples where the military did not hesitate to use its members as guinea pigs, such as the Porton Down experiments discussed previously, in which the test subjects who volunteered were not fully informed about the nature and the dangers of the substances they were exposed to. Stanley Mumford, a former volunteer, stated that research subjects were only 'given a broad idea and were told by the Medical Officer that there was no risk' (*Report of a Court of Inquiry*, 1953, p. 53).

Only a credible framework for military research can erase the suspicion that numerous service personnel and civilians might legitimately have about such practices. In order to establish such a trusting relationship, the use of technologies and medicines that might increase soldiers' capacities on the battlefield should occur only after a long experimental process. First, any military research aiming to increase soldiers' capacities should first pass the stage of animal testing. Of course, the use of animals in scientific research has been a subject of heated debate for many years. Those opposed to any kind of animal research believe that animal experimentation is cruel and unnecessary, regardless of its purpose or benefit. For these groups, there is no middle ground: they argue that all types of animal research should be banned once and for all. This point of view would have significant consequences for scientific research. Animal research plays a pivotal role in the discovery of new medicines and to give human beings a better quality of life and an increased life expectancy. However, this does not mean that scientists should be allowed to cause terrible and unnecessary suffering for animals. Some jurisdictions, like the United Kingdom and Germany, have adopted laws that favour animal welfare. If such experiments can demonstrate the value of the treatment being tested for potential diseases in soldiers and the absence of danger to their health, then protocols for human testing should be established.

At this point, and on the basis of Kantian ethics and an institutional duty of care, participation in these experiments should focus first and foremost on the consequences for the participants' immediate and long-term health, such as addiction; the potential of what the scientists are testing for the achievement of military objectives should never be the prime concern. The research should also analyse whether the medicines or technologies have a negative impact on soldiers' moral agency, more precisely their ability to distinguish between a lawful and an illegal order, as well as their capacity to maintain a total sense of responsibility for their actions – non-negotiable ethical guidelines that are so fundamental for the humanization of armed conflicts and the establishment of a just peace. If these criteria are not satisfied, a principle of precaution should prevail and scientists should return to the drawing board and modify the technologies or medicines.

It should also be clear that soldiers participate in the experiments on a voluntary basis, and that no pressure – whether explicit or implicit – should influence their willingness to test the technologies being developed. They should be informed about the purpose of the experiment and warned in advance of potential physical or mental discomforts that they may experience (as long as they are not unnecessary), and should also be free to bring the experiment to an end if they feel that the physical or mental discomfort has become unbearable. In order to make this theory a reality, superior commanding officers should never be present during the experiment, and a volunteer's refusal to pursue testing should not be mentioned in her or his military file, as scientists should not be allowed to report it to any member of the military authorities officially or unofficially.

The requirement for informed consent should be supplemented by the need for an ethical committee that has the responsibility to approve the testing on human subjects if it is deemed justified, and to supervise the experiments themselves. This committee would, of course, have to be independent from the military and the chain of command, in order to avoid any possible conflict of interests and to ensure that it does not prioritize military objectives, but rather the welfare and the rights of human volunteer subjects.

Such a process is by no means a novelty in the field of military research as it is very similar to the approach put forward during the Spanish American War, almost half a century before the establishment of the Nuremberg Code. At that time, US soldiers were threatened by yellow fever, and outbreaks were so frequent in the Caribbean that President Theodore Roosevelt asked the Army Surgeon General, General George M. Sternberg, to create a commission to study this disease and the ways in which it could be treated efficiently. The process of testing medicines on combatants was a milestone in military research at the time, even though it was attacked by antivivisectionists. During his address to the Congress of American Physicians in 1907, William Osler, a physician from Johns Hopkins Hospital and a key member of the Roosevelt commission, stated the importance of animal testing and the voluntary nature of participation in experiments, as follows:

> The limits of justifiable experimentation upon our fellow creatures are well and clearly defined. The final test of every new procedure, medical or surgical must be made on man, but never before it has been tried on animals. … For man absolute safety and full consent are the conditions which make such tests allowable. … Once this limit is transgressed the sacred cord which binds physician and patient snaps instantly. … Risk to the individual may be taken with his consent and full knowledge of the circumstances, as [has] been done in scores of cases, and we

cannot honour too highly the bravery of such men as the soldiers who voluntarily
submitted to the experiments on yellow fever in Cuba. (Osler, 1907, pp. 1–8)

Of course, from a strategic point of view, some might come to believe that these
procedures may be harmful to the secrecy that is necessary in order to accomplish
military objectives. In the case of a widespread disease like yellow fever, which
cannot be used intentionally by an armed force against its enemy, the need for
secrecy is not the same as it would be with the development of a sophisticated new
military device or medicine that might procure a strategic advantage. Moreover, it
would be a mistake to restrict secrecy simply to the need for military dominance.
Secrecy is also an essential component of the military's duty of care. While it
is difficult to defend a country or win a war if your enemy knows about your
strategy, armaments, technological capacities, or troop movement, a lack of secrecy
would also increase the vulnerability of soldiers, as it could allow the enemy to
know the Achilles heel of those it is fighting. From that perspective, protecting
military technological development is a moral obligation as a means to maximize
the military's duty of care. Thus being open with soldiers who test new technologies or medicines constitutes a risk. On the other hand, the twentieth-century
experiments discussed above prove that military secrecy has often been used as
a formidable way to cover up research that was so morally corrupt that it abused
human subjects and treated them merely as expendable goods. In other words,
while it can be deemed fundamental for preserving soldiers' life, it can also be a
way to treat them as guinea pigs.

However, one wonders whether a balance can be established between the need for
secrecy and the need for openness with those who are testing these new technologies
as a way to respect their welfare. Given the bond that unites members of the armed
forces, an insistence that their lives and their comrades' well-being depend on their
silence might be a strong incentive for them; the threat of dishonourable discharge
or judicial consequences would be also be a good compromise and would not be
an unacceptable burden from a moral perspective. Employees in other professional
organizations, such as civil servants, are also forbidden to disclose confidential
information they might encounter while performing their duties, under the threat
of significant consequences. For instance, in France, the law states that a serious
breach of confidentiality can lead to the immediate termination of an employee's
contract, and if it can be proved that the employee acted with intent to harm the
employer or to illegally profit from the disclosure of secret information, he or she
may be fined or imprisoned for up to three years.

Adoption of these guidelines would achieve numerous important goals. More
importantly, it would force the military to develop capacity-increasing technologies
that are not harmful to soldiers' health and their sense of responsibility and do not
hamper their right to ethical disobedience. Not only would soldiers be treated with
respect and as ends in themselves, but the guidelines would also help to preserve
the essential morality of warfare and would not run counter to the principles of *jus
post bellum*. Moreover, knowing that the development of technologies or medicines
respects these guidelines, members of the military would have no reason to fear

that their employer is acting out of pure paternalism, but rather would know that it is acting for their own good. Accordingly, waiving soldiers' consent for their use would not be as problematic as is the case at present.

On the other hand, if heads of state or commanders continue to use these technologies, in full knowledge of their consequences on soldiers' moral agency, this would definitely imply legal consequences for them according to the doctrine of command responsibility. This theory is intimately linked with the Japanese General Tomoyuki Yamashita, who was charged and sentenced to death by hanging for failing to prevent his men from committing war crimes during the Second World War. In their decision, the judges affirmed that those in positions of responsibility have an obligation to take all the 'appropriate measures' in their power 'to control the troops under [their] command for the prevention of the specified acts, which are violations of the law of war and which are likely to attend the occupation of hostile territory by an uncontrolled soldiery, and he may be charged with personal responsibility for his failure to take such measures when violations result' (In re Yamashita, 327 US 1 (1946)). In accordance with this logic, a commander can be prosecuted for dereliction of command responsibility insofar as it can be demonstrated that the individual was a superior, that he or she knew or had information that her or his subordinates were about to breach or in the process of breaching the laws or war, and that he or she did not take the appropriate measures to prevent the crimes from happening.

In fact, it would be possible to argue that military commanders who order the use of capacity-increasing technologies that would result in the perpetration of war crimes by their subordinates could share the blame for these violations if they knew or should have known that they were not entirely safe.. This sword of Damocles hanging over the heads of political leaders and military commanders would favour their use in a moral way. Of course, however, when faced with the reality of international law and its very often asymmetrical application between states, we must admit that this necessity may simply be wishful thinking.

Potential interactions between technologies or medicines also have to be analysed thoroughly during the experimental process. Although some may be unproblematic when used or taken separately, their combined use may cause problems for soldiers' health and moral agency. The same logic applies also to their use after an individual has suffered a previous medical issue. As mentioned previously, there are reasons to believe that Sergeant Bales's killing spree in Afghanistan may have been the result of mefloquine taken after he had previously suffered head injuries during his deployment in Iraq, a combination which, according to the drug manufacturer, may increase episodes of violence towards others. In order to respect the aforementioned moral principles of warfare, the military should avoid such potential risks. This creates an obligation on the part of the military to avoid treating all its members in the same way, and instead to introduce a more customized approach which aims to identify soldiers who may suffer from interactions between technologies or medicines. If there is such a risk for a handful of service personnel, the military should then have the responsibility to examine whether they could be exempted from using them without risk to their health and the safety of their comrades. If such a risk exists, the military should refuse their deployment and assign them other duties.

Of course, respecting these guidelines implies that the military should be seriously proactive in its research, in order to be ready for situations where it has to use untested or unapproved medicines on its members because of an unforeseen threat. This type of scenario led in the aftermath of the 1991 Gulf War to numerous criticisms and a shared feeling among service personnel and civilians that the military had used dangerous products, and is unlikely to happen again. The problem with such a situation concerns the uncertainty linked with the use of such medicines. Although they may actually increase soldiers' protection on the battlefield, they can also affect their health or moral agency – a risk that cannot be assessed until months or years after they have been used. Sometimes it is already too late. In the meantime, soldiers' health can be affected forever, and individuals can end up losing their moral agency and commit crimes for which responsibility may be very difficult to determine under current judicial principles. While the military's duty of care should be a given, the use of untested and unapproved technologies creates the possibility of its violation as well as the transgression of the moral principles that allow wars to be fought justly. This is why proactivity in military research and testing is so fundamental.

Overall, the use of capacity-increasing technologies remains morally necessary, so long as it is able to increase the safety of soldiers without posing a threat to their health or their moral agency on the battlefield. The distinction between acceptable and unacceptable technologies can be made only through a long process of experiment and testing. This is what this chapter has attempted to outline.

The use of technologies that may be to soldiers or play a negative role on the morality of warfare is often the result of insufficient testing or a lack of proactivity on the part of the military because it is forced to make emergency decisions. There is therefore a need to reconsider the ethical criteria for civilian research in the realm of the military. As mentioned, the main hurdle is the notion of consent. Allowing soldiers to consent to the use of technologies or medicines raises significant issues in relation to the military's duty of care for the reasons cited and, accordingly, forces us to think of a framework that would overlook consent while treating soldiers as ends in themselves and as autonomous moral agents.

Conclusion

Discussing capacity-increasing technologies in the military necessarily implies investigating the processes by which they are designed. After all, being able to distinguish which enhancements are morally acceptable from those that are not – if they do not treat soldiers as ends in themselves and if they threaten the principles of Just War Theory – can be achieved only through a long process of research and development. This shift is necessary because the military cannot ask the soldiers to consent to using capacity-increasing technologies, as the procedure might paradoxically turn out to be counterproductive and harmful to the institution's duty of care towards its members. For that reason, this chapter has discussed a series of criteria that might make the waiving of combatants' consent more acceptable than is currently the case.

Notes

1 In total, the United States conducted 1,054 nuclear tests (by official count) between 1945 and 1992, including 216 atmospheric tests.

2 From about August 1942 to about May 1943 experiments were conducted at the Dachau concentration camp, primarily on behalf of the German Luftwaffe, to determine the most effective means of treating persons who had been severely chilled or frozen. In one series of experiments, the subjects were forced to remain in a tank of iced water for periods of up to three hours. Extreme rigour developed in a short time. Numerous victims died in the course of the experiments.

3 From about March 1942 to about August 1942, experiments were conducted at Dachau on behalf of the German Luftwaffe to investigate the limits of human endurance and existence at extremely high altitudes. They were carried out in a low-pressure chamber in which the atmospheric conditions and pressures prevailing at high altitude (up to 68,000 feet) could be simulated. Many victims died as a result of these experiments, and others suffered grave injuries, torture, and ill treatment.

4 From about February 1942 to about April 1945 experiments were conducted at Dachau in order to investigate immunization against and treatment of malaria. Healthy inmates were infected by mosquitoes or by injections of extracts from the mucous glands of mosquitoes, and after contracting malaria, they were treated with various drugs to test their relative efficacy. Over 1,000 involuntary subjects were used in the experiments. Many of them died, and others suffered severe pain and permanent disability.

5 At various times between September 1939 and April 1945 experiments were conducted at Sachsenhausen, Natzweiler, and other concentration camps on behalf of the German armed forces to investigate the most effective treatment of wounds caused by Lost, a poisonous gas that is commonly known as mustard gas. Wounds were deliberately inflicted on the subjects and infected with Lost. Some of the subjects died as a result of these experiments, and others suffered intense pain and injury.

6 For instance, in October 2006, the Canadian Corporal Anthony Liwyj, a fully qualified vehicle technician, was asked to inspect the brake system on a Beaver Tail trailer. He noticed numerous problems, which were reported to his superior, who ordered him to correct them with air pressure only. Disagreeing with his sergeant, he insisted that the repair would be safer if it was done in another way. He was court-martialled for his refusal to obey the order. During his trial, Corporal Liwyj explained clearly and in great detail why, in his opinion, the brake adjustments he was asked to perform with only air pressure were unsafe, on the basis of his personal detailed observations of the situation and his personal knowledge and experience. Because the Canadian Forces Technical Orders for the trailer in question stated that the repair could be done by using only air pressure, the court concluded that the order was lawful, and consequently Corporal Liwyj was found guilty of disobedience (R. v. Corporal A. E. Liwyj, 2009 CM 3008). In another Canadian case, Master Seaman R.J. Middlemiss was found guilty of disobedience of a lawful command when, stationed at Colorado Springs in the United States in 2007, he declined an invitation to attend a mess dinner for which he would have to pay $35. Master Seaman Middlemiss argued that the forced attendance and payment at the mess dinner imposed on him an ideological conformity that violated his right to association. He also criticized the fact that it was an unplanned expense, and the implied enjoyable

participation in the mess dinner, contrary to his belief that, on the basis of personal experience, he would not have an enjoyable time. The court dismissed his claims on the basis that one of the common aspects for mess dinners consists in mandatory attendance (R. v. Master Seaman R.J. Middlemiss, 2009 CM 1002).

7   According to the US Food and Drug Administration, an investigational drug is a medicine used in clinical investigation. At the time when these drugs were administered to soldiers, neither had been studied in a formal clinical trial and no manufacturer was conducting studies.

8   This policy is currently implemented in the British and Canadian armed forces, both of which have adopted a voluntary approach for vaccination against anthrax (McManus et al., 2005, p. 1124).

9   After a senior officer gave further information about Private Dunn's health (he had spent two weeks in hospital with pneumonia until two days before the event), his sentence was remitted and replaced by ten years of hard labour.

10 The US Department of Defense rules state that 'unit officers and non-commissioned officers (NCOs) are specifically restricted from influencing the decisions of their subordinates to participate or not to participate as research subjects'. They also state that 'unit officers and senior NCOs in the chain of command are required to be absent during research subject solicitation and consenting activities' (quoted by Mehlman, 2015, p. 416). Finally, participants must be informed that 'participation is voluntary, that refusal to participate will involve no penalty or loss of benefits to which the subject is otherwise entitled, and that the subject may discontinue participation at any time' (US Army, 1989).

11 As echoed by Patrick Lin, Maxwell J. Mehlman, and Keith Abney, 'The form clearly states that consent is voluntary: "My decision to take Dexedrine", it reads, "is voluntary. I understand that I am not required to take the medication. Neither can I be punished if I decide not to take Dexedrine". But the form goes on to say: "However, should I choose not to take it under circumstances where its use appears indicated, I understand safety considerations may compel my commander, upon advice of the flight surgeon, to determine whether or not I should be considered unfit to fly a given mission". In other words, if you don't consent, you may not be allowed to fly. It's hard to imagine that pilots who have devoted so much time and effort to being able to fly in the military would refuse to take that drug' (Lin. Mehlman, and Abney, 2013, p. 74).

12 'Pervitin, la pilule de Goering', Arte Documentary 2015. www.youtube.com/watch?v= 1BHxWrZYlSI (last accessed 27 September 2017).

References

Advisory Committee on Human Radiation Experiments. 1995. *Final Report.* https://ehss.energy.gov/ohre/roadmap/achre/report.html (last accessed 7 October 2017).

Blum, William. 2006. *Rogue State: A Guide to the World's Only Superpower,* London: Zed Books.

Bower, Leah. 2003. 'Services: No Reason to Change Go-Pill Policy', *Stars and Stripes,* 23 February. https://www.stripes.com/news/services-no-reason-to-change-go-pill-policy-1.2322 (last accessed 8 October 2017).

Caldwell, G.G. et al. 1983. 'Mortality and Cancer Frequency Among Military Nuclear Test (SMOKY) Participants, 1957 through 1979', *Journal of the American Medical Association,* Vol. 250, No. 5, pp. 620–624.

Cockburn, Patrick. 1993. 'US Navy Tested Mustard Gas on its Own Sailors: In 1943 the Americans Used Humans in Secret Experiments', *The Independent*, 14 March.

Committee on Government Reform. 2000. *The Department of Defense Anthrax Vaccine Immunization Program: Unproven Force Protection*. United States Congress, House of Representatives. 106th Congress.

Doe v. Sullivan, 938 F.2d 1370 (US App. DC 1991).

Doe and Doe v. Sullivan, 756 F. Supp. 12 (USDC 1991).

Evans, Rob. 2004. 'The Past Porton Down can't Hide', *The Guardian*, 6 May.

Festing, Simon and Robin Wilkinson. 2007. 'The Ethics of Animal Research. Talking Point on the Use of Animals in Scientific Research', *European Molecular Biology OrganizationReport*, Vol. 8, No. 6, pp. 526–530.

Frisina, Michael. E. 2003. *Medical Ethics in Military Biomedical Research*. Textbooks of Military Medicine: Military Medical Ethics, Vol. 2. Office of The Surgeon General, Department of the Army.

Hart, Lianne. 2003. 'Use of Go Pills a Matter of Life and Death, Air Force Avows', *Los Angeles Times*, 17 January.

Hartle, Anthony E. 2004. *Moral Issues in Military Decision Making*. Lawrence, Kansas: University Press of Kansas.

In re Yamashita, 327 US 1 (1946).

Kamienski, Lukasz. 2016. *Shooting Up: A Short History of Drugs and War*. Oxford: Oxford University Press.

Khatchadourian, Raffi. 2012. 'Operation Delirium', *New Yorker*, 17 December.

Lin, Patrick, Maxwell J. Mehlman, and Keith Abney. 2013. *Enhanced Warfighters: Risk, Ethics, and Policy*. The Greenwall Foundation. http://ethics.calpoly.edu/Greenwall_report.pdf (last accessed 8 October 2017).

Martin, David S. 2012. 'Vets Feel Abandoned after Secret Drug Experiments', CNN, 1 March. http://edition.cnn.com/2012/03/01/health/human-test-subjects/index.html (last accessed 7 October 2017).

McManus, John et al. 2005. 'Informed Consent and Ethical Issues in Military Medical Research', *Academic Emergency Medicine*, Vol. 12, No. 11, pp. 1120–1126.

Mehlman, Maxwell J. 2015. 'Captain America and Iron Man: Biological, Genetic, and Psychological Enhancement and the Warrior Ethos'. In George R. Lucas (ed.), *Routledge Handbook of Military Ethics*. London and New York: Routledge, pp. 406–420

Mehlman, Maxwell J. and Stephanie Corley. 2014. 'A Framework for Military Bioethics', *Journal of Military Ethics*, Vol. 13, No. 4, pp. 331–349.

Milner, Claire Alida. 1997. 'Gulf War Guinea Pigs: Is Informed Consent Optional During War?', *Journal of Contemporary Health Law and Policy*, Vol. 13, No. 1, pp. 199–232.

Moreno, Jonathan. 2012. *Mind Wars: Brain Science and the Military in the 21st Century*. New York: Bellevue Literary Press.

Osler, William. 1907. 'The Historical Development and Relative Value of Laboratory and Clinical Methods in Diagnosis: The Evolution of the Idea of Experiment in Medicine', *Transactions of the Congress of American Physicians and Surgeons*, Vol. 7, pp. 1–8.

Parker v. Levy, 417 US 733, 743 (1974).

Pearce, N. et al. 1990. 'Follow-Up of the New Zealand Participants in British Atmospheric Nuclear Weapon Tests in the Pacific', *British Medical Journal*, No. 300, pp. 1161–1165.

Pierce, D.A. et al. 1996. 'Studies of the Mortality of Atomic Bomb Survivors', Report 12, Part 1, Cancer: 1950–1990', *Radiation Research*, Vol. 146, pp. 1–27.

Presidential Advisory Committee on Gulf War Veterans' Illness. 1997. *Final Report.* www.gulflink.osd.mil\gwvi/ (last accessed 8 October 2017).

R. v. Corporal A.E. Liwyj, 2009 CM 3008.

R. v. Master Seaman R.J. Middlemiss, 2009 CM 1002.

Raman, S., C.S. Dulberg, R.A. Spasoff, and T. Scott. 1987. 'Mortality among Canadian Military Personnel Exposed to Low-Dose Radiation', *Canadian Medical Association Journal*, No. 136, pp. 1051–1056.

Report of a Court of Inquiry. 1953. Reference AY.1030. Chemical Defence Experimental Establishment, Porton, Wiltshire, May.

Schmidt, Ulf. 2006. 'Cold War at Porton Down: Informed Consent in Britain's Biological and Chemical Warfare Experiments', *Cambridge Quarterly for Healthcare Ethics*, Vol. 15, No. 4, pp. 366–380

Sidel, Victor and Barry S. Levy. 2003. 'Physician-Soldier: A Moral Dilemma?'. In Edmund D. Pellegrino, Anthony E. Hartle, Edmund G. Howe, and Walter Reed (eds), *Military Medical Ethics*, Vol. 1. Textbooks of Military Medicine. Washington, DC: Department of the Army, pp. 293–312.

Simons, Marlise. 1993. 'Soviet Atom Test Used Thousands as Guinea Pigs, Archives Show', *New York Times*, 7 November.

Simpson, Doug. 2003. 'Air Force pushed Pilots to Take Amphetamines, Lawyer Says in Friendly Fire Case', *Detroit News*, 2 January.

Trials of War Criminals before the Nuremberg Military Tribunals under Control Council Law. 1949. Washington, DC: US Government Printing Office, No. 10.

US Army. 1989. Office of the Surgeon General, Reg. 15-2 I/11/89. https://mrmc-www.army.mil/docs/req/otsg15-2.pdf (last accessed 8 October 2017).

Watanabe, K.K., H.K. Kang, and N.A. Dalager. 1995. 'Cancer Mortality Risk among Military Participants of a 1958 Atmospheric Nuclear Weapons Test', *American Journal of Public Health*, Vol. 85, No. 4, pp. 523–527.

Wolfendale, Jessica and Steve Clarke. 2008. 'Paternalism, Consent, and the Use of Experimental Drugs in the Military', *Journal of Medicine and Philosophy*, Vol. 33, No. 4, pp. 337–355.

5

Permanent capacity-increasing technologies and the transhumanist danger

> Rights are the basis of our liberal democratic political order and key to contemporary thinking about moral and ethical issues.
>
> Francis Fukuyama, 2002

So far, this book has argued that the use of capacity-increasing technologies can be seen as a moral duty on the part of the military and should occur only if they do not pose a threat to soldiers' health and do not run counter to the principles of Just War Theory by depriving combatants of their moral agency and individual responsibility. An evaluation of permissible and impermissible techniques can be achieved effectively through a long and comprehensive testing process such as the one detailed in the previous chapter. However, this discussion has purposely ignored another important consideration in the debate: capacity-increasing technologies that give rise to permanent effects on soldiers. We can imagine a situation where these capacity-increasing technologies respect all the ethical guidelines previously discussed and, accordingly, are permissible on the basis of what has been argued so far. However, the discussion would be incomplete without considering the numerous moral problems associated with such technologies; these are fundamental problems and have been discussed only superficially in previous studies (Krishman, 2015; Pugliese, 2015).

Permanent capacity-increasing technologies are a constituent element of transhumanism, a movement aiming to transform the human condition by increasing its physical and psychological capacities. This global trend, which emerged at the turn of the new millennium,[1] has caused numerous debates among the scientific community about the potential of and dangers associated with this evolution in medical research, which is increasingly moving away from therapy and towards an ideal of enhancing the human body and mind. As the French philosopher Luc Ferry has argued, despite the increasing popularity of this debate within the intellectual community,[2] political leaders and citizens are, strangely, very quiet about it, even though this evolution will sooner rather than later transform our daily life from top to bottom (Ferry, 2016, p. 27). Civilians and members of the intellectual community have a duty to understand the legitimacy of these technologies because if they were ever developed and used, their implications would extend beyond the restricted world of the military and would affect everyone without exception. This chapter will first explain the nature of transhumanism, as well as its potential and dangers. It will argue that permanent capacity-increasing technologies under development by the military pose fundamental ethical problems in relation to notions of human egalitarianism and individual self-improvement. Secondly, it will argue that permanent capacity-increasing technologies may end up harming another pivotal principle of Western societies, namely, freedom: because of the secrecy that surrounds these technologies, there is a legitimate fear that individuals who benefit from them might see their freedom of movement hampered significantly by their state as a way to ensure that the technologies do not fall into the wrong hands.

What is transhumanism?

Developed two decades ago, transhumanism is defined by Nick Bostrom, one of its main proponents, as the idea 'that current human nature is provable through the use of applied science and other rational methods, which may make it possible to increase [the] human health-span, extend our intellectual and physical capacities, and give us increased control over our own mental states and moods' (Bostrom, 2005, pp. 202–203). To a very large extent, this new version of humanness is justified on the basis of its potential to extend the gains of secular humanism and the Enlightenment with the help of science. Humanist thinkers and philosophers of the Enlightenment always showed great optimism about human beings' ability to exceed the limits imposed by nature and were hopeful about the potential for an infinite perfectibility in humans. For example, Condorcet wrote in his famous *Sketch for a Historical Picture of the Progress of the Human Mind*:

> Our hopes for the future condition of the human race can be subsumed under three important heads: the abolition of inequality between nations, the progress of equality within each nation, and the true perfection of mankind. ... Is the human race to better itself, either by discoveries and sciences and the arts, and so in the means to individual welfare and general prosperity; or by progress in the principles of conduct or practical morality; or by a true perfection of the intellectual, moral, or physical faculties of man, an improvement which may result from a perfection either of the instruments used to heighten the intensity of these faculties and to direct their use or of the natural constitution of man? (Hyland, 2003, p. 29)

These humanist authors believed that this quest was possible mainly through the development of people's knowledge by means of public education and by their capacity to act according to their own free will instead of following the diktat of religion. But in the present day, individuals can go one step further thanks to the immense progress of science, which now enables them to engage in transforming their physical and intellectual capacities. This desire to develop attributes and abilities so far removed from the natural human condition is often referred to as transhumanism. What should we think of this idea? This hot ethical topic is being increasingly discussed by intellectuals, who have offered numerous in-depth analyses (Buchanan et al., 2001; Buchanan, 2011; Bostrom and Savulescu, 2009; Harris, 2007; Besnier, 2012; Vincent and Férone, 2011; Le Dévédec, 2015).

The transhumanist project is far more ambitious than the project that humanists and Enlightenment philosophers had in mind, as the improvement of the human condition does not depend solely upon education and cultural refinement. Rather, it lies in the use of scientific advances to overcome our biological limitations. With the help of science, transhumanists believe that we currently have at our disposal the tools that will enable us to transcend natural restrictions such as our lifespan, as well as our intellectual and psychological capacities, by redesigning the human being. Gene therapy and genetic engineering are integral to the transhumanist project, which may, ultimately, make humans the only species to take control of its evolution. It is generally agreed that the transhumanism will allow societies to increase people's fundamental rights to pursue happiness. It is in this light that Leon R. Kass, the former chairman of the President's Council on Bioethics, summarizes the reason behind this new trend:

> Our 'practice of happiness' – that's what [Thomas] Jefferson meant when he spoke of its 'pursuit' – could be made easier and more enjoyable than ever, because many of the natural obstacles to happiness – like the weaknesses of the body, the ravages of age, the vagaries of mood and temperament, and even the inherent unfairness of the genetic lottery – might be diminished or removed. (*Beyond Therapy*, 2003, p. vii)

Thus adopting a negative view of this topic would be a mistake as it would not do justice to the desirable outcomes that science can offer to humanity. However, as with the question of capacity-increasing technologies for members of the military, this does not mean that we should look at the issue through rose-coloured glasses. While transhumanism can actually be supported by moral principles, it would be a mistake to neglect its inherent ethical problems. If they were to be ignored, this movement might very well be, to use Francis Fukuyama's now famous remark, 'the most dangerous idea in the world' (Fukuyama, 2004). As with the ethics of super soldiers, a nuanced judgement must prevail. While many have come to believe that this quest is terrifying and a blasphemy against the order ordained by God or by nature, we cannot deny that it is established upon strong ethical principles that are in line with the founding ideals of Modernity. This is highlighted by Allen Buchanan's proposal to replace 'chance' with 'choice' in his discussion of transhumanist eugenism (Buchanan et al., 2001). Indeed, the progress of science can allow us to counteract the terrible natural lottery, which afflicts some individuals with serious cognitive deficiencies that will impair their future intellectual and social development: such unchosen gifts should not be accepted in light of the founding principles of Modernity, as they prevent the unlucky individuals from pursuing their own conception of the good life. With the help of technology, this source of human inequality may be corrected, as Nick Bostrom argues:

> The difference between the best times in life and the worst times is ultimately a difference in the way our atoms are arranged. In principle, that's amenable to technological innovation. This simple point is very important, because it shows

that there is no fundamental impossibility in enabling all of us to attain the good modes of being. (Quoted in Garreau, 2005, p. 242)

Through genetic modification, science might allow parents to have in vitro fertilization (IVF) of a foetus that will not run the risk of suffering from intellectual deficiencies. Although this possibility might frighten many, transhumanist thinkers argue that this form of eugenism is already an accepted reality in many societies on a much smaller scale. For example, Nick Bostrom and Rebecca Roache remind us that pregnant women are already encouraged to take folic acid supplements, which can positively affect the epigenetic expression of the baby, and that young girls receive vaccines against rubella to limit the risks of giving birth to babies with brain damage and other afflictions of congenital rubella syndrome (Bostrom and Roache, 2008, p. 18). Similarly, in the United States, 67 per cent of foetuses diagnosed with Down's syndrome are aborted. Accordingly, perhaps without even noticing it, some individuals have already practised this form of eugenism with no social stigma. Moreover, it should be noted that numerous pieces of legislation also allow a form of eugenism at the societial level. For instance, in Britain, the law permits couples who have a history of genetic disorders to select embryos without these genes for their IVF, and in Australia, parents are allowed to obtain pre-implantation genetic diagnoses to determine whether an implanted IVF embryo suffers from diseases like cystic fibrosis or haemophilia, as well as to determine its gender (Bostrom and Roache, 2008, p. 19).

Even though eugenism usually refers to the terrible activities practised by the Nazis, its proponents are very keen to point out that their approach is in no way similar to them.[3] Rather than aiming to eliminate those who are supposedly weak to ensure the purity of a pseudo-superior race, it is seen as a way to equalize the human condition and to counter the inherent inequalities of the natural lottery. Consequently, far from being socially discriminatory, this type of eugenism 'could potentially increase equality in society by enabling those with lower cognitive ability to function at the level that is closer to those with naturally high cognitive ability' (Bostrom and Roache, 2008, p. 16). In a certain way, this process is similar to the politics of difference proposed by Charles Taylor, discussed in the first chapter. Although individuals living in modern societies are supposed to be able to fulfil their own dreams and conceptions of the good life equally, social contingencies often make it more difficult for certain people to benefit from the same chances. To counterbalance such unequal opportunities, these societies have a duty to implement a politics of difference by treating people differently in order to ensure their right to equally pursue their freely chosen way of life without being victims of discrimination. Of course, social contingencies are not the only problems that can prevent individuals from enjoying equal opportunities in life. Natural infirmities and other genetic pathologies also play a constitutive role, as individuals suffering from intellectual disabilities are more likely than other people to earn lower incomes and to be the victims of social exclusion (Salkever, 1995). Although government programmes and initiatives can contribute to restoring balance between individuals, transhumanists believe that this problem can be solved in advance, and more efficiently, through genetic selection. As the Belgian philosopher Gilbert Hottois argues:

> Racist eugenism had no scientific foundation; it denied the essential equality of people; it did not respect the parents' autonomy: it was a state eugenism. The question of eugenism now has to be reconsidered in line with the equal dignity of all individuals and the desire to correct natural inequalities. Until now, distributive justice has been limited to the requirements of compensatory measures of the various inequalities: on the one hand, the inequalities associated with social lotteries (including the struggle against various forms of discrimination: those based on gender, ethnicity, race, and religion); and on the other hand, the inequalities caused by the natural lottery (health, lack of natural talent, etc.) without being able to intervene in its root cause. Until now, we have proceeded [in] an external manner, by giving financial compensation, free health care, special education, etc. The progress in genetics should give us a growing capacity to correct these inequalities before they appear, either by preventing them from occurring or by positive eugenism. If it was ever possible, we would have to evolve from a pure redistribution of social resources to a redistribution of natural resources (in other words, the genes) (Hottois, 2014, pp. 54–55, my translation).

Building on the Enlightenment notion of allowing everyone to freely pursue their own conception of happiness, transhumanist philosophers see eugenism as a moral duty, and as something that cannot be compared in any way to the Nazis' eugenics policies. It is from this perspective that Robert L. Sinsheimer argues: '[t]he old eugenics would have required a continual selection for breeding of the fit, and a culling of the unfit. The new eugenics would permit in principle the conversion of all the unfit to the highest genetic level' (Sinsheimer, 1992, p. 145).

This thesis has been subjected to numerous criticisms, as the line between selecting embryos without genetic disorders as a way to ensure greater equality (which is more associated with the capacity-restoring logic) and selecting ones with superior physical, psychological, and intellectual genetic codes can be very blurred. What might appear at first sight to be a modern duty in line with the Enlightenment tradition could eventually lead to a perverted path that we might come to regret. We would be putting our heads in the sand if we neglected the temptation to produce embryos with the perfect genes in order to create individuals with higher capacities than others. As Michael Sandel has pointed out, the breakthroughs in genetics could become an incentive to make individuals better than merely normal (Sandel, 2004). This eventuality is not a mere hypothesis. Sandel recalls that an Ivy League newspaper once published an advertisement seeking women who were at least five feet ten inches tall and athletic, with no family medical problems and with a combined SAT score of 1,400 or above who would be willing to sell an ovum in exchange for US$50,000. Another website claimed that it was offering ova from fashion models at prices ranging from US$15,000 to US$150,000. Moreover, in 1980, Robert Graham, a philanthropist whose mission was to improve the world's genetics by counteracting the rise of 'retrograde humans', opened one of America's first sperm banks, intending to collect sperm from former Nobel Prize winners and to create embryos with ova of women with superior IQs. Because of a lack of offers, he closed his clinic in 1999. This was an extreme case of enhancement eugenism. However, more commendable clinics may also be tempted by this logic, knowing that there will be a demand for such services. Sandel gives the example of the California Cryobank clinic:

> In contrast, California Cryobank, one of the world's leading sperm banks, is a for-profit company with no overt eugenic mission. Cappy Rothman, M.D., a co-founder of the firm, has nothing but disdain for Graham's eugenics, although the standards Cryobank imposes on the sperm it recruits are exacting. Cryobank has offices in Cambridge, Massachusetts, between Harvard and MIT, and in Palo Alto, California, near Stanford. It advertises for donors in campus newspapers (compensation up to $900 a month), and accepts less than five percent of the men who apply. Cryobank's marketing materials play up the prestigious source of its sperm. Its catalogue provides detailed information about the physical characteristics of each donor, along with his ethnic origin and college major. For an extra fee, prospective customers can buy the results of a test that assesses the donor's temperament and character type. Rothman reports that Cryobank's ideal sperm donor is six feet tall, with brown eyes, blond hair, and dimples, and has a college degree – not because the company wants to propagate those traits, but because those are the traits his customers want: 'If our customers wanted high school dropouts, we would give them high school dropouts.' (Sandel, 2004)

This version of eugenism, which no longer promotes a desirable equality among humans, but rather an unfairness between those with enhanced genetics and those without them, is not associated with the modern project, as it might result in creating inequalities. There is, in fact, a danger of creating discrimination – which is precisely what transhumanism seeks to eliminate – between two classes of citizens: those who have been given special physical features unnaturally and those who have had to rely on the natural lottery. While the latter are currently regarded as the citizens with normal abilities, in the long run they run the risk of being in a situation of disadvantage in relation to the former. Thus even though the motivations of transhumanists in favour of eugenism are nothing like those of the Nazis, their quest could potentially create the same situation. For example, the Nazis deliberately targeted those they deemed inferior, and they were consequently segregated and discriminated against. As already mentioned, contemporary eugenism instead aims to targeting unborn individuals who could be victims of unequal chances in their later life because of natural contingencies, by correcting the potential sources of future discrimination so that they can fully and freely pursue their conception of the good life. However, an unregulated eugenism that allows certain individuals to enjoy intellectual or physical potentialities beyond the normal – and in view of the current trends in the commercialization of genetic background, we can presume that these individuals will come from rich families – runs the risk of establishing a similar dichotomy between superior human beings and others who are defined as weak. This potential danger is central to proponents of transhumanism like Bostrom, who writes:

> We can imagine scenarios where such inequities grow much larger thanks to genetic interventions that only the rich can afford, adding genetic advantages to

the environmental advantages already benefiting privileged children. We could even speculate about the members of the privileged stratum of society eventually enhancing themselves and their offspring to a point where the human species, for many practical purposes, splits into two or more species that have little in common except a shared evolutionary history. The genetically privileged might become ageless, healthy, super-geniuses of flawless physical beauty, who are graced with a sparkling wit and a disarmingly self-deprecating sense of humor, radiating warmth, empathetic charm, and relaxed confidence. The non-privileged would remain as people are today but perhaps deprived of some of their self-respect and suffering occasional bouts of envy. The mobility between the lower and the upper classes might disappear, and a child born to poor parents, lacking genetic enhancements, might find it impossible to successfully compete against the super-children of the rich. Even if no discrimination or exploitation of the lower class occurred, there is still something disturbing about the prospect of a society with such extreme inequalities. (Bostrom, 2003, p. 502)

This situation is seen by certain authors as a danger to intersubjective solidarity. As has been discussed by Sandel, if the prospect of transhumanism ever becomes generalized in our societies, it could erode our sense of responsibility towards others who have not been not favoured by the natural lottery (Sandel, 2007, p. 89). More precisely, knowing that we could have been born with an intellectual impairment or with a gene that increases our chances of developing cancer or some other disease is, for Sandel, a constitutive reason for the natural solidarity we owe to the disfavoured. There is, in this sense, a close connection between solidarity and natural giftedness. Because the latter is contingent on an uncontrollable element, it creates an obligation of humility about our own natural talents and a willingness to share the fruits of our good fortune with those who have not have had the chance to inherit these gifts. To highlight his point, Sandel uses the case of personal insurance:

> Since people do not know whether or when various ills will befall them, they pool their risk by buying health insurance and life insurance. As life plays itself out, the healthy wind up subsidizing the unhealthy, and those who live to a ripe old age wind up subsidizing the families of those who die before their time. The result is mutuality by inadvertence. Even without a sense of mutual obligation, people pool their risks and resources, and share one another's fate. (Sandel, 2007, pp. 89–90)

With genetic engineering, this natural obligation could simply be smashed to pieces, as individuals who are able to benefit would know that their chances of developing a potentially deadly disease are non-existent and might opt out of the pool, which would dramatically increase the premiums of those who might still be victims of the natural lottery. For this reason, Sandel is of the opinion that genetic enhancement 'would make it harder to foster the moral sentiments that social solidarity requires' (Sandel, 2007, pp. 90–91). On the contrary, he writes that

> If genetic engineering enabled us to override the results of the genetic lottery, to replace chance with choice, the gifted character of human powers and achievements would recede, and with it, perhaps, our capacity to see ourselves as sharing a common fate. The successful would become even more likely than they are now to

view themselves as self-made and self-sufficient, and hence wholly responsible for their success. Those at the bottom of society would be viewed not as disadvantaged, and so worthy of a measure of compensation, but as simply unfit, and so worthy of eugenic repair. The meritocracy, less chastened by chance, would become harder, less forgiving. As perfect genetic knowledge would end the simulacrum of solidarity in insurance markets, perfect genetic control would erode the actual solidarity that arises when men and women reflect on the contingency of their talents and fortunes. (Sandel, 2007, pp. 91–92)

Francis Fukuyama also discusses this risk by stating that the natural lottery, although profoundly unjust, nonetheless contributes to making us sensitive to the misfortune of our fellow citizens.

Today, many bright and successful young people believe that they owe their success to accidents of birth and upbringing but for which their lives might have taken a very different course. They feel themselves, in other words, to be lucky that they are capable of feeling sympathy for people who are less lucky than they. But to the extent that they become 'children of choice' who have been genetically selected by their parents for certain characteristics, they may come to believe increasingly that their success is a matter not just of luck but of good choices and planning on the part of their parents, and hence something deserved. They will look, think, act and perhaps even feel differently from those who were not similarly chosen, and may come in time to think of themselves as different kinds of creature. They may, in short, feel themselves to be aristocrats, and unlike aristocrats of old, their claim to better birth will be rooted in nature and not convention. (Fukuyama, 2002, p. 157)

In fact, this dichotomy between the haves and the have-nots could trigger a new form of natural discrimination that would be more difficult to fight than the conventional forms of discrimination against particular groups of people (McKibben, 2004). In the past, individuals have been deprived of certain fundamental rights, such as those who were forced to become slaves in antiquity as a result of a conquest. For other groups, their subordination was deemed to be justified by their inferior or servile nature (such as Amerindians, women, or the poor), but it nonetheless relied on pseudo-scientific beliefs and must therefore be considered a conventional form of discrimination. However, with eugenism, the differentiation between humans could take a truly different path by relying for the first time in history on a scientific assessment of human nature, which could be more difficult to challenge than one based on conventional criteria.

Of course, as previously mentioned, it is worthless to adopt a binary perspective regarding the transhumanist revolution. Its inherent problems cannot be ignored but, on the other hand, its moral justifications can hardly be discarded as worthless and unfounded. As with the question of capacity-increasing technologies in the military, it may be best to reach a balanced solution through specific regulations. Two regulatory principles may be invoked. First, to ensure that this scientific progress remains an outgrowth of the modern principle of equality, state policies could limit the techniques of eugenism to medical conditions that might create discrimination and prevent individuals from enjoying equal rights to pursue happiness. In other words, they should remain only within the realm of therapy. This should accordingly restrict the use of techniques that might allow certain people to benefit from enhanced physical, intellectual, or psychological capacities. Second, and this is a pivotal argument for proponents of transhumanism, if societies were to accept moving from therapy to enhancement, these techniques should be made equally available to everyone (Bostrom, 2005, p. 203) through state subsidies or by providing them freely to children of poor parents (Bostrom, 2003, p. 503), and not only to a handful of individuals. This would avoid the creation of two classes of citizens.

The use of permanent capacity-increasing technologies in the military and the risk of creating first-class citizens

It seems clear that the US Army is moving towards capacity-increasing technologies that will have permanent effects on soldiers. Numerous reports have explicitly stated the potential benefits of this new approach for the military as a way to enhance soldiers' health and improve their performance on the battlefield (National Research Council, 2001; Melson, 2004, JASON, 2010), and the belief is also fuelled by reports that the military has shown an interest in changing soldiers' cellular and genetic structures to enable them to run longer distances, to survive longer without food, or to be able to consume foods that are not digestible today, such as grass (Shachtman, 2007). DARPA has also financially supported research to develop blood transfusions with genetically modified cells that can neutralize lethal biological toxins. According to one of this study's senior scientists, soldiers would be able to enjoy 'long-lasting reserves of antioxin antibodies' (Conor, 2014). This research is in line with many of DARPA's projects, such as the Advanced Tools for Mammalian Genome Engineering, which aims to add a forty-seventh chromosome to human bodies as a vector platform for inserting bio-alterations and wholesale genetic 'improvements' into our DNA (DARPA, 2013), or its Living Foundries programme, which attempts to change the genetic makeup of soldiers to make them stronger, more resistant, and more resilient to biological threats (DARPA, 2012). Finally, in view of the dangerous impact of sleep deprivation on soldiers, the US Air Force has financed a report in order to understand how this problem might be solved through genetic engineering (JASON, 2008). As lack of sleep disrupts more than seven hundred genes, the hope is to find a way to reprogram them in order to allow soldiers to carry out missions despite sleep deprivation.

Other nations also seem to be moving in this direction, even though they are not so open as the United States about what they are doing. We only have a glimpse into such information, but reports have suggested that scientists in China have managed to create 'super beagle dogs' which are stronger than normal dogs because the myostatin gene has been deleted at the embryo stage. When asked about the potential of this research, a member of the project said that 'it was possible that humans could be genetically modified, like the beagles, to make stronger athletes or better soldiers' (Heilpern, 2015).

This scientific progress is clearly along the lines of transhumanism and shows that permanent enhancement in the military is more than a mere dream and cannot be ignored, even though these developments may be realized only in the long term. However, there is a need to evaluate the desirability of permanent enhancements from an ethical standpoint before they are used on battlefields.

Of course, on the basis of what has been argued so far, enhancements should not be tolerated if they are to be treated as mere means to an end. For example, capacity-increasing technologies that require embryonic modifications to deliberately create a pre-specified class of future warriors (Ford and Glymour, 2014, p. 46) would not qualify as morally acceptable, as they would dictate these individuals' way of life without allowing them any autonomy. Scientific research regarding these types of gene manipulation does not seem to have reached this point, but they nonetheless remain conceivable in the future. Genetic modifications of soldiers would also be ethically unacceptable if they were to negatively alter soldiers' health, deprive them of their obligation to disobey illegal and immoral orders, or negatively affect the moral principles that deal with the fair termination phase of war. But let us assume that they were respectful of these criteria after having been tested in an ethical fashion and would not create a situation of riskless warfare for the members of armed forces benefiting from them. Should we allow their use or not? The answer to this question seems to be negative in view of the aforementioned problems that are inherent to the transhumanist movement. The main reasons for this are linked with the fact that such enhancements will most probably be restricted to combatants and will never be available to civilians. Indeed, as mentioned in the previous chapter, the military has moral grounds to ensure that its strategic developments remain secret in order to minimize soldiers' exposure to death or injury. It would therefore be counterproductive for a state to democratize these innovations as they might easily become available to its enemies, thereby harming its duty of care towards its members. The fact that these enhancement techniques might be given to only a limited number of individuals entails numerous risks.

First, membership of the military is for most soldiers simply a career which, in the US armed forces, may last up to fifteen years for enlisted personnel and up to eleven years for officers. In the British naval service officers, depending on their specialism and commission type, can leave after six, eight, twelve, or sixteen years. The majority of army and Royal Air Force (RAF) officers, by contrast, serve on longer contracts. The RAF has clear outflow peaks at the ages of thirty-eight, forty-four, and fifty-five, reflecting its various retirement options, whereas the army's main outflow is at fifty-five, the normal retirement age. In France, the average retirement age for members of the military is 43.8 years (51.3 for officers, 45.8 for non-commissioned officers, and 32.2 for enlisted personnel): significantly younger than for civil servants of the Hexagon, who retire, on average, at 58.8 years. This means that when they retire, French soldiers are still young enough to begin a new career, a possibility which is also fostered by the fact that their average monthly pension (which in some cases is not indexed or is indexed at a lower percentage rate than inflation) is significantly lower than those of civil servants (€1,484 compared with €1,957). The same imbalance affects service personnel from the US military in comparison with federal civil servants: the average annual federal pension is US$32,824, but US$22,492 for veterans (Cauchon and D'Ambrosio,

2012). Because of this, many veterans choose – either because they are too young to retire permanently or because their pension is not enough to maintain a decent living – to return to the job market.

It is true that many veterans face numerous hurdles in their transition from military life to the civilian workforce. The main problem is their struggle to translate their military skills into civilian job experience, which creates a significant barrier to employment. As explained by the White House in 2015:

> Many high-demand, good-paying jobs like paramedics, truck drivers, nurses, and welders, require either a national certification or state occupational license to be hired, and currently our national and state systems make it very difficult for service members and veterans to obtain these civilian certifications and licenses that directly translate to their military training. Often, service members and veterans are required to repeat education or training to receive these occupational credentials, even though much, and in some cases, all of their military training and experience overlaps with credential training requirements. Moreover, employers, many with significant needs for skilled workers, are left waiting for these military members to complete these, oftentimes lengthy, credentialing training programs – programs that many veterans could have taught themselves. (White House, 2015)

To help retired service personnel acquire the necessary credentials to join the civilian workforce, the US government has implemented measures to make it easier for them to gain the necessary skills and qualifications. These types of measures are common in numerous societies. For instance, in Massachusetts, individuals who need a new skill to get a job can benefit from a state programme that waives the requirement to seek work while they are enrolled full time in agency-approved training, while receiving unemployment insurance payments for an extended period of twenty-six weeks if they remain enrolled in their programme. France has similar programmes that offer free professional training for unemployed individuals while allowing them to retain their unemployment insurance payments. These programmes are in line with the aforementioned politics of difference and aim to help disadvantaged or underprivileged people to meet the minimum workforce requirements that will allow them to avoid social discrimination. Through this equalization process, individuals are then able to compete fairly with other job seekers by showing employers that they have the necessary skills, experience, and personal qualities to be hired. On the basis of the principle of facilitating equality among unemployed people, there is, in this sense, nothing reprehensible about such policies that conform to the politics of difference.

However, what if certain people have extraordinary unnatural advantages in comparison with others, such as former soldiers who benefit from permanent capacity-increasing technologies? For instance, they may be enabled by such technologies to fully focus on their task without being distracted, work extended hours without feeling tired, or show above average physical strength or a boosted IQ close to the maximum score of 161. Employers would surely not remain insensitive to these qualities and might be more willing to hire these veterans than civilians who do not enjoy the same advantages. In this case, we would be moving from a situation of equality to one based on unnatural inequalities, as these enhancements would be reserved for only a small minority of individuals; this is a fundamental problem of transhumanism, and is admitted by own proponents.

Of course, the distribution of intellectual, physical, and psychological talents between individuals is naturally uneven, and this also affects the possibilities of fulfilling our respective dreams and living our lives the way we want to. For example, children who are not as gifted as other students in science or mathematics have less chance of becoming economists, doctors, or astronauts. The same logic applies for individuals who wish to become professional cyclists but who have a low VO_2 max (maximal oxygen consumption rate) or a weak resistance to intense pain. However, these natural inequalities are not insurmountable and can be overcome in various ways, for example through hard work and determination – probably one of the main sources of personal motivation that gives value to our most cherished life objectives and the desire to surpass ourselves.

This way of achieving our goals is at the heart of contemporary notions of success and recognition. As discussed by Charles Taylor, since the Enlightenment our societies have no longer been organized around an idea of unequal honours.[4] Within democracy, recognition now rests on universalist and egalitarian principles (Taylor, 1994, p. 27). Everyone is now theoretically entitled to enjoy recognition by others: this is a defining feature of how we can shape our own personal identity. However, while it used to be an *a priori* feature attributed solely on the basis of our belonging to a specific social class, we now need to build this recognition ourselves through ingenuity, courage, or dedication. This is why personal successes and performances are so important, as they help us acquire social recognition. There is now growing pressure on individuals to rise above others, and despite our unequal natural capacities, this possibility is largely open to all of us. This process implies comparing ourselves with others in order to determine our own limits and understand what we need to do to become better.

The reader will perhaps allow me to discuss some personal experiences that will help to highlight this point. Some of my students have greater capacities to summarize a text well or to write deeper and more brilliant papers than others. I am always impressed when B-grade students ask me to explain what separates them from A-grade students. I am attentive to this request, eager to point out their flaws, to encourage them to work harder, and to make greater efforts in the hope that they will manage to close the gap and become as accomplished as some of their peers. Similarly, as a former amateur athlete with average physical abilities, I will always remember my first race in 2009 – a winter pentathlon in Quebec City involving cycling, jogging, cross-country skiing, speed skating, and snow-shoe running – in which I started with very high expectations. However, I did not have the means to realize my ambitions as I quickly fell behind and ended up in a humiliating seventy-fifth position. This smack in the face served as an incentive to understand where I stood in comparison with other athletes and to analyse what I needed to do to achieve victory. The failure led me to train harder and more intelligently. With this lesson in mind, I was back the next year on the starting line and finished in third position, for which I was praised by my friends and other rivals. Whenever I look at the bronze medal I am always amazed that a person like me who was not born with superior physical skills can achieve some kind of recognition with effort and dedication.[5] I am sure that those of my students who received an A grade for a paper after receiving an average grade in their previous paper had the same feelings that I had when I stood on the podium in February 2010. The same logic applies to an individual who wishes to obtain a highly competitive job. If he or she realizes after the first examination that he is not in the top 5 per cent of the class, he or she may be motivated by this relative failure to identify and correct what caused this, by studying either harder or more intelligently. Eventually, the dedication may pay off and he or she may become the best-qualified candidate for the job. Thus the challenge of our modern society is not to produce a situation of pure equality between humans, but rather a form of fairness. This means that it is necessary to encourage people to innovate and pursue their personal development, and this will allow some of them to rise above the rest. As explained by Fritz Allhoff et al.,

> There are good reasons to think that we want some gap to exist, for example, to provide incentives for innovations, in order to move up the economic ladder, and to allow flexibility in a workforce to fill vacancies and perform a wide range of tasks. At least some competition seems to be desirable, especially when resources to be allocated are limited or scarce and when compared to the historically unsuccessful alternative of the state attempting to equalize the welfare of its citizens. (Allhoff et al., 2010, p. 17)

However, this idea of fairness depends in large measure on the need to create equal opportunities between individuals, such as good-quality, free, compulsory education or accessible public libraries that will allow, for instance, the poorest child to gain the means to compete on an equal playing field with others. Once these minimum requirements are established, it is up to individuals to benefit from them and to prove their merit.

With these examples in mind, let us imagine that individuals now have to compete against people benefiting from unnatural enhanced capacities – in this case, former members of the military with permanent genetic modifications – that others cannot equal even with the highest degree of dedication or effort. They might simply become 'dinosaurs in a hypercompetitive world' (Allhoff et al., 2010, p. 17) and the natural reality could therefore become unbalanced; two types of citizens would be created and, as argued by Sandel, this might decrease the degree of solidarity between them, as well as devalue the whole concept of realizing one's goals and of exceeding what one believes can be achieved through hard work. This point should not be taken lightly, as for many of us having goals, being able to accomplish some of them, and having reasonable expectations of fulfilling other objectives are what make life meaningful and worth leading. If this possibility was taken away or hindered, life would lose a part of its meaning. It would also mean that achieving recognition would become more difficult than it is today and would favour only a handful of individuals. This would be unfair, as it would facilitate granting honours and higher social status to individuals on grounds other than merit. Such a situation would have certain similarities with the actions of cheats or dopers who circumvent the usual equalitarian path to social recognition. This would impair the modern conception of the politics of recognition, a fundamental human need that contributes to defining who we are as human beings.

In return, those benefiting from unnatural enhancements might develop an idea of superiority over those without enhancements, creating a depreciated image of their value and potential. This danger should not be taken lightly. As mentioned by Taylor, our identity as human beings is affected not only by the recognition given by others, but also by their '*mis*recognition' (Taylor, 1994, p. 25). If the unenhanced are seen by their enhanced counterparts as members of an inferior group, they may, like women in patriarchal societies and black people in societies dominated by white people, end up developing a demeaning and oppressed image of themselves as insignificant and worthless individuals (Taylor, 1994, pp. 25–26). For this reason, Taylor argues that 'misrecognition shows not just a lack of due respect. It can inflict a grievous wound, saddling its victims with a crippling self-hatred' (Taylor, 1994, p. 26).

Permanent military transhumanism and its implications for its beneficiaries

While the prospect of transhumanism can be seen as advantageous for members of the military, it must nonetheless be pointed out that permanent capacity-increasing technologies might also entail significant problems. The most important is clearly the lack of freedom that its recipients might have to experience. As with any other capacity-increasing technologies of the past, it is reasonable to think that the military would resort to genetic modifications simply in order to gain advantages over its foes. This would transform these enhancement techniques into secret weapons that need to be hidden from foreign eyes as a matter of national security. Needless to say, it would be extremely difficult for the military to protect the secrecy of bio-engineered weapons used by its former members after they are discharged. While enrolled individuals might expect that their freedom will be hindered during their service, they would expect to regain their full freedom, such as being able to travel anywhere they want, at the end of the contract. Now, it is fairly easy to imagine that a former soldier who is benefiting from permanent capacity-increasing technologies would not want to travel in an explicitly or implicitly enemy state where he or she could be kidnapped and subject to medical experimentation that allowed this nation to fully apprehend the technology and to transfer it to its own combatants. As Nicholas G. Evans and Jonathan Moreno put it, in such a situation 'A soldier's body could become a security risk, in the same way as the loss of an unmanned aerial vehicle over enemy territory incurs the risk of unintended technology transfer' (Evans and Moreno, 2014, p. 6). In the case of non-human enhanced technologies, such as drones or weapons, the military has traditional methods of protection, most notably by restricting its sales to untrusted states. If we transposed these methods to bio-engineered retired soldiers, it might very well entail the military restricting their freedom in the same way, primarily by denying them the right to leave their home country. Such a decision would, of course, be contrary to another main component of our modern world, namely the right to travel, which is integral to various domestic constitutions and international conventions, such as Article 13 of the Universal Declaration of Human Rights.

Of course, it is true that this right – like any other fundamental right in liberal societies – is not absolute and may be hindered on reasonable grounds, such as the need for national security. The United States provides a good example of this, since the president is allowed to deny or to revoke US citizens' passports for matters of foreign policy or national security. Historically, this restrictive policy has been applied against individuals suspected of communist sympathies. More recently it has also been implemented against individuals who have left or tried to leave their home country in order to fight alongside the forces of the Islamic State in the Middle East.

From that standpoint, we could argue that it is reasonable to prevent former members of the military who are benefiting from permanent capacity-increasing technologies from traveling in order to avoid losing information intimately connected with matters of national defence and security. If we agree with this position, it nonetheless leaves the door open to a unique criticism. Unlike the bans on freedom of movement imposed on former communists or youth who decide to fight the Jihad, a ban on former permanently enhanced soldiers' freedom of movement would be ongoing because genetic modifications cannot be suppressed. History provides us with numerous examples of this, such as Sinn Fein's politicians Gerry Adams and Martin McGuinness, who after being involved for decades in a terrorist organization responsible for acts of violence, the Irish Republican Army, as a way to promote the option of a united Ireland, are now fully committed to achieving this goal through democratic means. The same logic also applies to Nelson Mandela, the father of the South African civil rights movement, who spent the first part of his adult life behind bars for his involvement in attacks against civilians. These two examples show that people are free to choose and to revise their respective conceptions of the 'good life' and can change from being individuals whose freedom reasonably deserves to be restricted to individuals who no longer pose a harm to the lives and safety of others. When this is the case, it becomes unreasonable to restrict their freedom.

However, in contrast to people who have the possibility of abandoning a path that is contrary to national security, permanently enhanced retired soldiers would never be able to regain the right to unrestricted freedom. Of course, one could argue that such a restriction would be acceptable if all the future consequences – especially the permanent restriction on freedom of movement – were to be explained before their biological transformation. However, this would be unable to fully take into account the fact that people's decisions are not binding for ever. It is assumed in our modern world that all individuals have an equal right to choose and revise their conception of the good life at any given time. For instance, a soldier could very well decide at the time of their enlistment – let us say at twenty years old – that travelling is an unnecessary and expensive luxury that they would never need, and so receiving a genetic transformation would never affect their future freedom of movement. However, there is no guarantee that this belief will remain the same for ever. During their tenure in the military, they might very well be deployed to numerous countries and develop a taste for travelling and adventure which they would like to pursue after retirement. The initial decision taken at the age of twenty could then make it impossible for them to fully enjoy their new conception of the good life. It would also have implications for their future employment after retirement from the military, as it would prevent them from getting a job that involved travel to international conferences or to meet clients abroad. There is therefore a major difference between the logic behind the current restrictions on civilians' freedom of movement and one that could affect permanently enhanced warfighters. Unlike the former, the latter would never be able to regain this fundamental right because of a choice that cannot be reversed.

It is also necessary to consider the collateral effects of this choice on third parties, namely the children of permanently enhanced soldiers. Since the genetic modifications would be passed on to them, they would experience not only from their advantages, but also from their disadvantages. As the bearers of a secret military technology that should never fall into enemy hands, they too might conceivably, like their parents, see their freedom of movement restricted. In this situation, how would it be possible to morally justify a government decision that would harm a pivotal element of people's freedom as the result of a choice made by somebody else? Against their will, these children would see their capacity to pursue happiness severely restricted in comparison to that of children who did not inherit genetic modifications, notably by being forbidden to move from one place to another and work in economic spheres that required them to travel. Paradoxically, this possibility would lead to a form of discrimination by creating two classes of citizens: those who can enjoy certain rights and those who cannot, either because of their own irreversible decision or because of one taken by their parents.

Conclusion

Our ability to benefit from an equal capacity to gain recognition – which, as mentioned, can be promoted through a politics of difference for those who face social or physical struggles – is a gain of Modernity over ancient hierarchical societies. Alongside our ability to act according to our own free will and the necessity to treat others as ends and not as means, equal opportunities to pursue our own conceptions of happiness constitute one of the main factors upon which human dignity and our liberal societies are established. Undermining them would undoubtedly set off a dynamic with still unknown consequences. However, before the world's militaries begin to proceed with these biotechnological innovations, it is necessary to discuss the possible negative consequences that might follow from them. Do we want to live in a world where individuals have, like doped athletes, an unnatural edge over their fellow citizens, a situation that might alter people's motivation to rise above others solely through their determination and their desire to exceed what they believe are their limits? Would the use of these enhancements on a small scale eventually create natural discrimination between people? Would restricted permanent enhancement harm our sense of solidarity towards others who have been victims of the natural lottery? Would it tend to unreasonably limit people's right to pursue their own freely chosen conception of the good life? These questions deserve global attention because, unlike the non-permanent capacity-increasing technologies discussed in the previous chapters, they could affect all of us, and future generations, if they are ever developed and used.

Notes

1 The first report on the subject was written in 2002 and published a year later as entitled *Converging Technologies for Improving Human Performance: Nanotechnology, Biotechnology, Information Technology and Cognitive Science* (Roco and Sims Bainbridge, 2003) and had a considerable impact.
2 Companies have also shown an interest in this debate, notably by creating in 2008 the Singularity University, an institution financed by Google in Silicon Valley.
3 The Nazis did not invent this heinous form of eugenism. They simply followed a path established before they gained power. As Francis Fukuyama appropriately recalls, 'In the late nineteenth and early twentieth centuries, state-sponsored eugenics programs attracted surprisingly broad support, not just from right-wing racists and social Darwinists, but from such progressives as the Fabian socialists Beatrice and Sidney Webb and George Bernard Shaw, the communists J.B.S. Haldane and J.D. Bernal, and the feminist and birth-control proponent Margaret Sanger. The United States and other Western countries passed eugenics laws permitting the state to involuntarily sterilize people deemed "imbeciles", while encouraging people with desirable characteristics to have as many children as possible. In the words of Justice Oliver Wendell Holmes, "We want people who are healthy, good-natured, emotionally stable, sympathetic, and smart. We do not want idiots, imbeciles, paupers, and criminals"' (Fukuyama, 2002, p. 85).
4 In the military world, the creation in 1802 of the French military academy of Saint-Cyr by Napoleon represents a good example. Until the French Revolution, the officer class was restricted to members of the aristocracy; Napoleon democratized it and opened it to all French men solely on the basis of merit.
5 Those who argue that performance-enhancing drugs in sporting events should be legalized do not fully grasp the importance of this dimension of sport, as these artificial methods of achieving our goals would create a shortcut that would eliminate human endeavour and dedication (Savulescu, Foddy, and Clayton, 2004).

References

Allhoff, Fritz et al. 2010. 'Ethics and Human Enhancement: 25 Questions and Answers', *Studies in Ethics, Law, and Technology*, Vol. 4, No. 1, pp. 1–39.
Besnier, Jean-Michel. 2012. *Demain les posthumains*. Paris: Fayard.
Beyond Therapy: Biotechnology and the Pursuit of Happiness: A Report of the President's Council on Bioethics. 2003. Washington, DC: Dana Press.
Bostrom, Nick. 2003. 'Human Genetic Enhancements: A Transhumanist Perspective', *Journal of Value Inquiry*, Vol. 37, No. 4, pp. 493–506.
Bostrom, Nick. 2005. 'In Defense of Posthuman Dignity', *Bioethics*, Vol. 19, No. 3, pp. 202–214.
Bostrom, Nick and Rebecca Roache. 2008. 'Ethical Issues in Human Enhancement'. In Jesper Ryberg, Thomas Petersen, and Clark Wolf (eds), *New Waves in Applied Ethics*.

London: Palgrave Macmillan, pp. 120–152. http://www.nickbostrom.com/ethics/human-enhancement.pdf (last accessed 11 October 2017).

Bostrom, Nick and Julian Savulescu. 2009. *Human Enhancement.* Oxford: Oxford University Press.

Buchanan, Allen. 2011. *Better than Human: The Promise and Perils of Enhancing Ourselves.* Oxford: Oxford University Press.

Buchanan, Allen et al. 2001. *From Chance to Choice: Genetics and Justice.* Cambridge: Cambridge University Press.

Cauchon, Dennis and Paul D'Ambrosio. 2012. 'Some Federal Pensions Pay Handsome Rewards', *USA Today*, 15 August.

Conor, Steve. 2014. 'Future Soldiers Could be Protected against Germ Warfare by Genetically Modified Blood Cells', *The Independent*, 1 July.

DARPA (Defense Advanced Research Projects Agency). 2012. 'Living Foundries'. Biological Technologies Office. www.darpa.mil/Our_Work/BTO/Programs/Living_Foundries.aspx (last accessed 11 October 2017).

DARPA (Defense Advanced Research Projects Agency). 2013. 13.B Small Business Technology Transfer (STTR) Program Proposal Submission Instructions. www.acq.osd.mil/osbp/sbir/solicitations/sttr2013B/darpa13B.htm (last accessed 11 October 2017).

Evans, Nicholas G. and Jonathan D. Moreno. 2014. 'Yesterday's War; Tomorrow's Technology: Peer Commentary on "Ethical, Legal, Social and Policy Issues in the Use of Genomic Technologies by the US Military"', *Journal of Law and Biosciences*, Vol. 3, No. 1, pp. 1–6.

Ferry, Luc. 2016. *La révolution transhumaniste: comment la technomédecine et l'ubérisation du monde vont bouleverser nos vie.* Paris: Plon.

Ford, Kenneth and Clark Glymour. 2014. 'The Enhanced Warfighter', *Bulletin of Atomic Scientists,* Vol. 70, No. 1, pp. 43–53.

Fukuyama, Francis. 2002. *Our Posthuman Future: Consequences of the Biotechnology Revolution*, London: Profile Books.

Fukuyama, Francis. 2004. 'Transhumanism', *Foreign Policy*, September–October. http://foreignpolicy.com/2009/10/23/transhumanism/ (last accessed 11 October 2017).

Garreau, Joel. 2005. *Radical Evolution.* New York: Doubleday.

Harris, John. 2007. *Enhancing Revolution: The Ethical Case for Making People Better.* Princeton: Princeton University Press.

Heilpern, Will. 2015. 'Super-Strong, Genetically-Engineered Dogs: Could they Cure Parkinson's Disease?', CNN, 28 October. http://edition.cnn.com/2015/10/28/world/china-mutant-dogs-genetic-engineering/index.html (last accessed 11 October 2017).

Hottois, Gilbert. 2014. *Le transhumanisme est-il un humanisme?* Brussels: Éditions de l'Académie Royale de Belgique.

Hyland, Paul (ed.). 2003. *The Enlightenment: A Sourcebook and Reader.* London: Routledge.

Krishman, Armin. 2015. 'Enhanced Warfighters as Private Military Contractors'. In Jai Galliott and Mianna Lotz (eds), *Super Soldiers: The Ethical, Legal and Social Implications.* Farnham, UK: Ashgate, pp. 65–80.

JASON. 2008. *Human Performance.* https://www.fas.org/irp/agency/dod/jason/human.pdf (last accessed 11 October 2017).

JASON. 2010. *The $100 Genome: Implications for the DoD.* http://fas.org/irp/agency/dod/jason/hundred.pdf (last accessed 11 October 2017).

Le Dévédec, Nicolas. 2015. *La société de l'amélioration: la perfectibilité humaine, des Lumières au transhumanisme.* Montreal: Liber.

McKibben, Bill. 2004. *Enough: Staying Human in an Engineered Age.* New York: Henry Holt and Co.

Melson, Ashley R. 2004. 'Bioterrorism, Biodefense, and Biotechnology in the Military: A Comparative Analysis of Legal and Ethical Issues in the Research, Development, and Use of Biotechnological Products on American and British Soldiers', *Albany Law Journal of Science and Technology*, Vol. 14, No. 2, pp. 1–44.

National Research Council. 2001. *Opportunities in Biotechnology for Future Army Applications.* Washington, DC: National Academies Press.

Pugliese, Joseph. 2015. 'On Human and Medicine: When is a Soldier not a Soldier?' In Jai Galliott and Mianna Lotz (eds), *Super Soldiers: The Ethical, Legal and Social Implications.* Farnham, UK: Ashgate, pp. 25–36.

Roco, Mihail C. and William Sims Bainbridge. 2003. *Converging Technologies for Improving Human Performance: Nanotechnology, Biotechnology, Information Technology and Cognitive Science.* Dordrecht: Kluwer Academic Publishers.

Salkever, David S. 1995. 'Updated Estimates of Earnings Benefits from Reduced Exposure of Children to Environmental Lead', *Environmental Research*, Vol. 70, No. 1, pp. 1–6.

Sandel, Michael J. 2004. 'The Case against Perfection', *Atlantic Monthly*, April. www.theatlantic.com/magazine/archive/2004/04/the-case-against-perfection/302927/ (last accessed 11 October 2017).

Sandel, Michael J. 2007. *The Case against Perfection: Ethics in the Age of Genetic Engineering.* Cambridge, Massachusetts: Belknap Press.

Savulescu, Julian, Bennett Foddy, and M. Clayton. 2004. 'Why we should Allow Performance Enhancing Drugs in Sports', *British Journal of Sports Medicine*, Vol. 38, No. 6, pp. 666–670.

Shachtman, Noah. 2007. 'Kill Proof, Animal-Esque Soldiers: DARPA Goal', *Wired*, 7 August.

Sinsheimer, Robert L. 1992. 'The Prospect of Designed Genetic Change' In Ruth F. Chadwick (ed.), *Ethics, Reproduction, and Genetic Control.* London and New York: Routledge, pp. 136–146.

Taylor, Charles. 1994. 'The Politics of Recognition'. In Amy Gutmann (ed.), *Multiculturalism.* Princeton: Princeton University Press, pp. 25–73.

Vincent, Jean-Didier and Geneviève Férone. 2011. *Bienvenue en transhumanie.* Paris: Grasset.

White House. 2015. 'States Step Up to Help Veterans Get Back to Work'. www.whitehouse.gov/joiningforces/veterans-back-to-work (last accessed 11 October 2017).

Conclusion

Our judgement about many issues appears at first to be fairly straightforward: they are often resolved as completely acceptable or unacceptable, as if we were unable to appreciate the necessary nuances surrounding them. This Manichaean assessment is pointless. With regard to analysing war and justifying the killing of other human beings, such black-and-white logic is counterproductive in the scope of its ethical implications. While it is easy to fall into this trap when discussing the development of super soldiers, it would be a mistake to do so, as we have seen that significant moral arguments can at the same time justify and restrict their development and use. Therefore caution and a balanced appraisal are necessary.

As already mentioned, capacity-increasing technologies can be seen as a necessary outcome of the military's duty towards its members, as they can increase soldiers' chances of survival and can also contribute to an enhanced morality of warfare, a principle that seems to be gaining ground in many countries such as the United Kingdom. As has been argued, the asymmetrical distribution of these technologies is not in itself morally problematic. It is certainly true that their use may significantly reduce the risk of some combatants being harmed or killed, but this is not a reason to object to their use, as some authors have argued. Insofar as they do not totally remove their beneficiaries from these risks, they do not in any way challenge the traditional understanding of warfare as a duel. Moreover, we cannot ignore the fact that capacity-increasing technologies can also play a positive role in the humanization of conflicts by helping soldiers to deal with stress, fatigue, anger, and many other similar emotions that have contributed to the perpetration of war crimes in the past. It is possible to argue that this type of technology might play a positive role in the ethics of warfare by limiting the risk of such atrocities being committed. However, these ethical outcomes cannot alone justify their use, because capacity-increasing technologies can also lead to moral drift, such as treating humans as mere means and weakening the principles of Just War Theory. The main fear associated with these technologies is the infringement upon soldiers' moral agency, which could lead to a situation where it would be impossible to assign them criminal and legal responsibility for their crimes. In light of the principles of *jus post bellum*, this judicial vacuum is of course a matter of concern. Taking this clash into consideration, this book has tried to determine the extent to which these technologies are acceptable as well as to identify the ways in which they should be developed so that they can be implemented in a morally justifiable manner, particularly by focusing on the ways in which they are tested and developed.

While this framework could be implemented today, the distant prospect of permanent capacity-increasing technologies is, however, more troubling, as it threatens to undermine one of the main philosophical principles of Modernity. Indeed, military transhumanism runs the risk of providing combatants with undue advantages over civilians, thereby creating an insurmountable chasm between the two groups in terms of opportunities. This could have significant social consequences in terms of solidarity, the constant desire of human beings to overcome their limits, and our capacity to freely choose our own paths to happiness. These developments will most certainly be one of the next steps in military enhancement. In view of this, human societies must carefully examine the possible repercussions. If they ignore this evolution and agree to allow some individuals to benefit from permanent superhuman features, one of the main pillars of the modern world will collapse, and we will regress into the ancient world that our ancestors fought to get rid of. In short, capacity-increasing technologies should to be allowed in the military only under the conditions discussed in this book, more specifically as long as they (a) do not create a situation of riskless warfare for some combatants; (b) respect the moral rules of warfare; (c) do not instrumentalize soldiers when the technologies are developed and used for the sake of another end; (d) are used only when we are certain that they will not create responsibility gaps; and (e) have only temporary effects on soldiers. Providing that these conditions are met, capacity-increasing technologies are not especially problematic from an ethical standpoint any more than other weaponry currently allowed on the battlefield.

However, we must realize that an increase in the physical and cognitive capacities of soldiers is only one of the many dramatic changes that will forever change the way our wars are fought; the development of super soldiers may be only the tip of the iceberg with regard to the role of technology in war. Thus we cannot ignore the challenges associated with the development of autonomous weapons, which have until now existed only in Hollywood movies and science fiction. As in the case of super soldiers, fiction is about to become a stinging reality that will not only affect the military world. Indeed, the move towards the autonomization of machines will soon become a feature of our daily lives, along with all of its consequences. For instance, Amazon is now famous for its highly publicized drone delivery system, and several companies, such as Uber and Google, have already started to test self-driving trucks. It has been estimated that when the use of autonomous vehicles peaks in several decades' time, truck driving will be a job of the past.[1] Many fear that autonomous technologies will eventually invade other sectors of the economy, to a point at which robots will replace humans in the workforce. While this fear is somehow exaggerated in the sense that there is no direct correlation between the use of robots and increased unemployment,[2] the evolution towards robotic autonomization should nonetheless prompt us to think about what our future will look like.

However, the prospect of autonomous weapons in the military is far more problematic, as it would mark the first time in human history that machines could decide a person's fate on their own, without any human intervention. It is no wonder that many campaigns have been initiated, by various groups, to ban autonomous weapons from the battlefields of tomorrow.[3] However, this ban stands no realistic chance of ever being implemented, given the enthusiasm of powerful nations for these weapons. For instance, the Pentagon has recently announced that it will invest US$18 billion in research and development of such technologies. The prospect may be nearer than people think, as General Paul J. Selva, the vice-chairman of the Joint Chiefs of Staff, has stated that the United States is only a decade away from having them (Rosenberg and Markoff, 2016). Reports also indicate that Russia and China are developing autonomous killing robots (Muoio, 2015). Although the United Kingdom has not yet announced any plans to develop such weapons in the future, it should be noted that in 2015 it refused to participate in an international ban on their development (Bowcott, 2015). Of course, we cannot ignore the dangers associated with these killing machines deciding who should live and who should die. The more general implications behind the negation of Isaac Asimov's famous 'Three Laws of Robotics' (Asimov, 1950) are also relevant here: (1) a robot may not injure a human being or, through inaction, allow a human being to come to harm; (2) a robot must obey the orders given to it by human beings except where such orders would conflict with the first law; and (3) a robot must protect its own existence as long as such protection does not conflict with the first and second laws.

Nevertheless, just as in the case of capacity-increasing technologies, adopting a binary perspective on the development of these weapons may turn out to be very unproductive. Of course, allotting power over the life and death of human beings to autonomous machines is more than troubling. That said, it would be a mistake to ignore the fact that they also have the potential to transform warfare for the better. The arguments put forward in this book regarding the moral use of capacity-increasing technologies can help us to determine the ethical potential and moral flaws of autonomous weapons. Like the logic behind the use of capacity-increasing technology, the desire to replace conventional human soldiers with autonomous robots is clearly justified under B.J. Strawser's principle of unnecessary risk, mentioned earlier. This is argued by Alex Leveringhaus, for whom autonomous robots are ethically justified because the military and the state owe a duty of care towards their members:

> Soldiers are, under certain circumstances, expected to make the ultimate sacrifice. However, militaries are obliged, within reasonable grounds, to minimise risks to their service personnel. Sending soldiers into battle with inadequate or faulty equipment would certainly be a violation of duties of care. … autonomous weapons reduce physical risks to their operators because they are uninhabited. Uninhabited weapons offer the greatest possible level of protection. As such, it might be obligatory to deploy autonomous weapons rather than soldiers, if possible. (Leveringhaus, 2016, p. 61)

Moreover, as with the question of capacity-increasing technologies, others have argued that autonomous armed robots might actually perform better than conventional soldiers with regard to respecting the principles of Just War Theory (Arkin, 2010). Although soldiers have always been the main components of the military, he argues, they are at the same time its weakest actors. Not only can they fail to conquer a state if they are demoralized, tired, or afraid of dying, but they have also been integral to numerous violations of international humanitarian law throughout history. Indeed, when confronted by the 'fog of war', anger that clouds their judgement, or the fear of being killed, they are sometimes either unable or unwilling to differentiate between friends and enemies or between civilians and combatants, which tends to create situations in which the rules of war will be violated. Robots have, on the other hand, certain features that might actually allow a more humane treatment of those who should not be harmed during warfare. As Arkin reminds us, armed autonomous robots are not driven by their fear of dying and, therefore, are exempt from the behaviour of 'shooting first and asking questions later', which very often explains unjustified killings. Moreover, contrary to the logic of even the most well-trained soldiers, robots can be designed in such a way that their actions will not be influenced by emotions such as anger and frustration, which are often the triggering factors for criminal behaviour (Arkin, 2010, pp. 333–334). In this sense, much like capacity-increasing technologies, autonomous robots present us with a real opportunity to improve compliance with the laws of war, which are too often violated by human beings.

On the other hand, despite their ethical advantages, autonomous weapons face the same types of criticism as capacity-increasing technologies with regard to criminal responsibility and their potential harmful consequences on *jus post bellum*. There are strong reasons to believe that autonomous robots may not be able to apply one of the most important requirements of *jus in bello*, discrimination between combatants and non-combatants. As has been mentioned, it is expected that individuals who violate this rule by attacking non-combatants will be prosecuted after war ends as a way to discourage future similar actions and to provide a form of justice for the families of the victims. This is possible only when the culprit is able to take full responsibility for her or his crime. This would be impossible if, at the time of the misdemeanour, he or she was unable to distinguish between good and evil because of a mental disability or as a result of being in a state of involuntary intoxication. This form of criminal non-responsibility is currently the only form of 'responsibility gap' that is tolerated in many jurisdictions in relation to the assignment of liability for a crime. As has been discussed, without proper assessment, capacity-increasing technologies in the military may open the door to an increased number of cases in which soldiers could legitimately use this type of defence, thereby making it even more difficult to prosecute war crimes. The use of autonomous weapons raises the same type of concern. Because soldiers operating them are essentially unarmed and the decision to kill is made without any human interaction, we can only wonder who would be blamed in a situation in which an innocent civilian was terminated by one of these machines. As Leveringhaus states, 'indiscriminate bombing could lead to widespread resentment amongst the civilian population in the enemy state, [thereby] undermining the prospects for long-term peace' (Leveringhaus, 2016, p. 18). This bitterness can be minimized when those responsible are tried and sentenced at the end of the conflict, but this can hardly be done when such indiscriminate bombings are carried out by autonomous machines. Just as is the case with untested capacity-increasing technologies, the autonomization of weapons can render obsolete the necessary prosecution of those responsible for war crimes in favour of a judicial vacuum whereby unlawful acts would remain unprosecuted. Those who favour the utilization of autonomous weapons cannot afford the luxury of trying to find ways to eliminate, or at the very least minimize as far as possible, the risk that these machines might create responsibility gaps. This implies the need for a thorough assessment of these weapons prior to their deployment on the battlefield in order to make sure that they will not run the risk of going against the moral rules of warfare and harming the principles of *jus post bellum*.

In sum, it has become obvious that with the development of super soldiers and autonomous robots, tomorrow's wars will become more akin to what we see in comic books or on the big screen. However, this does not mean that the principles of war ethics ought to be reviewed. On the contrary, war ethicists currently have all the necessary tools to analyse whether these new forms of weaponry are morally acceptable or not. This is what this book has proved. It has also called upon these specialists to seriously challenge these new developments without falling into the trap of excessive paranoia or absolute refusal to allow their deployment. Such a stand would do no good and would be like fighting a desperate and useless rearguard battle. The forces of politics, realism, and the military-industrial complex are too powerful to be ignored. The pen will be stronger than the sword only if political philosophy takes a balanced approach to the future nature of warfare by trying to assess the value of these technologies as well as their inherent problems.

As argued in this book, despite the flaws of capacity-increasing technologies, it would be a mistake to analyse them solely in black or white terms. While we have to be careful with them for valid ethical reasons, we cannot ignore the fact that they can have significant moral implications as well. Only a balanced perspective can help us establish normative rules to govern the emergence of these technologies and their not-so-distant deployment on the battlefields of tomorrow. I simply hope that this book has laid the foundations for a debate that we will inevitably have to face in the coming years.

Notes

1 In 2014, there were 3.1 million truck drivers in the United States, which represented 2% of the total employment of the country (Balakrishnan, 2017).
2 The countries that have the most robots (Japan, South Korea, Germany, and Sweden) are also those that have the lowest levels of unemployment.
3 For instance, this prohibition was discussed at the United Nations in 2014 and 2015.

References

Arkin, Ronald C. 2010. 'The Case for Ethical Autonomy in Unmanned Systems', *Journal of Military Ethics*, Vol. 9, No. 4, pp. 332–341.

Asimov, Isaac. 1950. *I, Robot*. New York: Gnome Press.

Balakrishnan, Anita. 2017. 'Self-Driving Cars Could Cost America's Professional Drivers up to 25,000 Jobs a Month, Goldman Sachs Says', CNBC, 22 May. https://www.cnbc.com/2017/05/22/goldman-sachs-analysis-of-autonomous-vehicle-job-loss.html (last accessed 12 October 2017).

Bowcott, Owen. 2015. 'UK Opposes International Ban on Developing "Killer Robots"', *The Guardian*, 13 April.

Leveringhaus, Alex. 2016. *Ethics and Autonomous Weapons*. Oxford: Palgrave Macmillan.

Muoio, Danielle. 2015. 'Russia and China are Building Highly Autonomous Killer Robots', *Business Insider*, 15 December.

Rosenberg, Matthew and John Markoff. 2016. 'The Pentagon's "Terminator Conundrum": Robots that Could Kill on their Own', *New York Times*, 25 October.

Index

EU authorised representative for GPSR:
Easy Access System Europe, Mustamäe tee 50,
10621 Tallinn, Estonia
gpsr.requests@easproject.com

www.ingramcontent.com/pod-product-compliance
Lightning Source LLC
Chambersburg PA
CBHW052013270326
41929CB00015B/2901